STUDIES IN ALGEBRAIC GEOMETRY

MAA STUDIES IN MATHEMATICS

Published by
THE MATHEMATICAL ASSOCIATION OF AMERICA

Studies in Mathematics

The Mathematical Association of America

Volume 1: STUDIES IN MODERN ANALYSIS
edited by R. C. Buck

Volume 2: STUDIES IN MODERN ALGEBRA
edited by A. A. Albert

Volume 3: STUDIES IN REAL AND COMPLEX ANALYSIS
edited by I. I. Hirschman, Jr.

Volume 4: STUDIES IN GLOBAL GEOMETRY AND ANALYSIS
edited by S. S. Chern

Volume 5: STUDIES IN MODERN TOPOLOGY
edited by P. J. Hilton

Volume 6: STUDIES IN NUMBER THEORY
edited by W. J. LeVeque

Volume 7: STUDIES IN APPLIED MATHEMATICS
edited by A. H. Taub

Volume 8: STUDIES IN MODEL THEORY
edited by M. D. Morley

Volume 9: STUDIES IN ALGEBRAIC LOGIC
edited by Aubert Daigneault

Volume 10: STUDIES IN OPTIMIZATION
edited by G. B. Dantzig and B. C. Eaves

Volume 11: STUDIES IN GRAPH THEORY, PART I
edited by D. R. Fulkerson

Volume 12: STUDIES IN GRAPH THEORY, PART II
edited by D. R. Fulkerson

Volume 13: STUDIES IN HARMONIC ANALYSIS
edited by J. M. Ash

Volume 14: STUDIES IN ORDINARY DIFFERENTIAL EQUATIONS
edited by Jack Hale

Volume 15: STUDIES IN MATHEMATICAL BIOLOGY, PART I
edited by S. A. Levin

Volume 16: STUDIES IN MATHEMATICAL BIOLOGY, PART II
edited by S. A. Levin

Volume 17: STUDIES IN COMBINATORICS
edited by G.-C. Rota

Volume 18: STUDIES IN PROBABILITY THEORY
edited by Murray Rosenblatt

Volume 19: STUDIES IN STATISTICS
edited by R. V. Hogg

Volume 20: STUDIES IN ALGEBRAIC GEOMETRY
edited by A. Seidenberg

Steven L. Kleiman
Massachusetts Institute of Technology

Jack Ohm
Louisiana State University, Baton Rouge

Maxwell Rosenlicht
University of California, Berkeley

A. Seidenberg
University of California, Berkeley

B. L. van der Waerden
Mathematical Institute, Zurich

Studies in Mathematics

Volume 20

STUDIES IN ALGEBRAIC GEOMETRY

A. Seidenberg, editor
University of California, Berkeley

Published and distributed by
The Mathematical Association of America

Library of Congress Catalog Card Number 80-81041

Complete Set ISBN 0-88385-100-8
Vol. 20 ISBN 0-88385-120-2

Printed in the United States of America

Current printing (last digit):

10 9 8 7 6 5 4 3 2 1

PREFACE

It is no secret that Algebraic Geometry has a vast literature, largely indigestible. If the student does not know this already, he soon will upon approaching it. The papers found in modern research journals are so written that the reader will not be able to make headway unless his mind is already well supplied with the notions being employed. Indeed, our graduate courses are intended precisely to equip the student with some background; but at best this will only help him in getting at a small part of the literature.

If one wishes, one can find some good reasons for the current situation, but probably one cannot do much about it anyway. At most the situation can be occasionally alleviated by expository articles. Below, a few such articles in Algebraic Geometry are offered.

Here are some remarks on the individual papers.

Professor Rosenlicht's paper on linear algebraic groups is almost completely self-contained. Anyone familiar with a few basic topological terms and with undergraduate algebra should have no difficulty in following the paper. Instead of dealing with just one linear transformation, a case with which the reader may be familiar, one deals with a group of them. Even with the elementary means to which Rosenlicht restricts himself he is able to establish some basic results, including the Lie-Kolchin theorem on solvable linear algebraic groups.

Professor van der Waerden's paper on the connectedness theorem and intersection multiplicity is also rather self-contained. It starts with a review of the basic knowledge needed; this part is easily accessible to the student, but the theorems on connectedness and multiplicity could only be fitted with difficulty into a first course in Algebraic Geometry. These are topics which have had vast extensions and have been developed with highly technical means; but still the student will be grateful for van der Waerden's simple treatment, which yields the basic forms of some far-reaching results.

Professor Ohm's paper is not self-contained: the reader will have to know some Homological Algebra. As algebra develops, its advances have a tendency to invade Algebraic Geometry: in the case of Homological Algebra the invasion is gaining momentum and has been doing so for years. Ohm's exposition consists in showing how Homological Algebra is applied in Algebraic Geometry, in particular to the question of how many polynomials are needed for the generation of the prime ideal of an affine space curve. The terms from Algebraic Geometry are defined and, in general, Ohm supplies the right amount of detail for an expository article. For the reader who knows the fundamentals of Homological Algebra the paper will be most helpful in making a link with Algebraic Geometry.

Professor Kleiman's paper is not directly expository, but it is indirectly so via a historical treatment of intersection multiplicity, especially with reference to a problem on conics. The ultimate source is the "problem of Apollonius," to construct a circle tangent to three given ones; Jacob Steiner gave the problem its modern form in 1848 by replacing circles by conics in the projective plane over the complex numbers and asking how many solutions there are in general. Before this question could be answered, it had to undergo considerable formalization; in particular, instead of conics, one spoke of complete conics—a *complete conic* is, roughly, the notion of a point conic and its dual. Beyond that there emerged two distinct formalizations, that of Halphen and that of Study; in 1938 van der Waerden accepted Study's view, whereas in 1940 Severi accepted Halphen's. Meanwhile interest in the original problem had lapsed; it has been given a definitive answer only within the past few years. The 1894 work of Shubert on similar problems in higher dimensional space still remains to be put in a form satisfactory to ourselves. Kleiman's paper will give the reader a healthy reminder of the vast difference between working on some recent aspect of a problem, however fascinating and significant this may be, and answering the original motivating question.

A. SEIDENBERG

CONTENTS

INITIAL RESULTS IN THE THEORY OF LINEAR ALGEBRAIC GROUPS

Maxwell Rosenlicht

The purpose of this article is to give an elementary exposition of a number of fundamental results about linear algebraic groups and at the same time some flavor of the general subject, all in a fairly self-contained way. Among the attractions of this field of mathematics are its interconnections with several other fields. Simple arguments typical of different disciplines combine to give transparent proofs of important facts about linear algebraic groups. Thus the notion of connectedness in topology combines with elementary arguments from group theory and linear algebra to prove the Lie-Kolchin theorem. Similarly, easily stated results about linear algebraic groups have unexpected applications to other fields, but this is an aspect we do not enter into here. Confining ourselves to the use of a few very elementary tools, not going beyond elementary facts from linear algebra and the Hilbert basis theorem for polynomial ideals, we obtain several results that are both attractive and substantial.

1. THE ZARISKI TOPOLOGY

We fix an algebraically closed field k, whose elements will be used to coordinatize the geometric objects of our concern. Thus the basic objects we deal with will be, as sets, subsets of some "affine space" k^n, where n is a positive integer. For any n, k^n will have a natural topology, the so-called *Zariski topology*, which is the smallest topology on k^n in which the set of zeros of any polynomial function in n variables with coefficients in k is closed. That is, taking $X_1, \ldots, X_n$ to be indeterminates over k, we postulate that for each $F \in k[X_1, \ldots, X_n]$ the set $\{(x_1, \ldots, x_n) \in k^n : F(x_1, \ldots, x_n) = 0\}$ be closed. More precisely, we define a subset of k^n to be *closed* if and only if there exists $\{F_\alpha\}_{\alpha \in A} \subset k[X_1, \ldots, X_n]$ such that the subset is the set of zeros of $\{F_\alpha\}_{\alpha \in A}$, that is, the set $\{(x_1, \ldots, x_n) \in k^n : F_\alpha(x_1, \ldots, x_n) = 0 \text{ for all } \alpha \in A\}$. If $\{F_\alpha\}_{\alpha \in A}, \{F_\beta\}_{\beta \in B}, \{F_\gamma\}_{\gamma \in C}, \ldots$ are subsets of $k[X_1, \ldots, X_n]$, the intersection of their sets of zeros in k^n is the set of zeros of $\{F_\alpha\}_{\alpha \in A} \cup \{F_\beta\}_{\beta \in B} \cup \{F_\gamma\}_{\gamma \in C} \cup \cdots$, while the union of the zeros of $\{F_\alpha\}_{\alpha \in A}$ and the zeros of $\{F_\beta\}_{\beta \in B}$ is the set of zeros of $\{F_\alpha F_\beta\}_{(\alpha, \beta) \in A \times B}$. The set of zeros of 1 is $\emptyset$ and of 0 is k^n. Thus there is a topology of k^n in which the closed sets are exactly as described. This topology is not Hausdorff if $n > 0$, for two nonempty open subsets of k^n necessarily have a nonempty intersection; this follows from the fact that any nonempty open subset of k^n contains all $(x_1, \ldots, x_n) \in k^n$ such that $F(x_1, \ldots, x_n) \neq 0$, for some nonzero $F \in k[X_1, \ldots, X_n]$, and the fact that the product of two nonzero elements of $k[X_1, \ldots, X_n]$ is also nonzero, therefore not annulled by at least some points of k^n. However, in the Zariski topology of k^n at least points are closed, for $(x_1, \ldots, x_n) \in k^n$ is the set of zeros of $\{X_1 - x_1, \ldots, X_n - x_n\}$.

A nonempty closed subset of k^n is called *irreducible* if it is not the union of two proper nonempty closed subsets. Thus k^n itself is irreducible, since any two of its nonempty open subsets have a nonempty intersection. If an irreducible closed subset of k^n is contained in the union of two other closed subsets, it is contained in one of them.

Together with the Zariski topology on k^n we can consider the induced topology on any subset of k^n. We shall do this quite often, especially for subsets that are themselves closed, or open, or the intersection of a closed set and an open set.

THEOREM 1: *The space k^n, and any subspace of k^n, is noetherian; that is, a strictly descending sequence of closed sets is necessarily finite. Like any noetherian space, it is quasicompact; that is, any open covering has a finite subcovering. Any closed subset of a noetherian space is uniquely expressible as the union of a finite number of irreducible closed subsets none of which contains another.*

To be specific, let us prove the first statement for k^n itself. Let $V_1 \supset V_2 \supset V_3 \supset \cdots$ be an infinite sequence of distinct closed subsets of k^n. If $\mathcal{I}(V_i)$ is the set of all elements of $k[X_1, \ldots, X_n]$ which vanish on V_i, then $\mathcal{I}(V_i)$ is an ideal in $k[X_1, \ldots, X_n]$ and V_i is the set of zeros of $\mathcal{I}(V_i)$. Since the V_i's are distinct, so are the $\mathcal{I}(V_i)$'s, and we get the strictly ascending chain of ideals $\mathcal{I}(V_1) \subset \mathcal{I}(V_2) \subset \cdots$, a contradiction of the Hilbert basis theorem. The quasicompactness of a noetherian space is clear. Next, if a noetherian space has a closed subset that is not the union of a finite number of irreducible closed subsets, it must have a minimal such closed subset V. For such a minimal V, if $V = V_1 \cup V_2$, where V_1 and V_2 are closed, at least one of V_1, V_2 is not the union of a finite number of irreducible closed subsets, so from the minimality of V we deduce that V_1 or V_2 equals V, and hence V is irreducible or empty. Thus any closed subset of V is the union of a finite number of irreducible closed subsets, hence, throwing some away if necessary, the union of a finite number of irreducible closed subsets none of which contains another. The uniqueness part is easy, since if $V_1, \ldots, V_r, W_1, \ldots, W_s$ are irreducible closed subsets such that $V_1 \cup \cdots \cup V_r = W_1 \cup \cdots \cup W_s$, then for each $i = 1, \ldots, r$ we have $V_i \subset W_1 \cup \cdots \cup W_s$, and thus $V_i \subset W_j$ for some $j = 1, \ldots, s$; similarly $W_j \subset V_m$ for some $m = 1, \ldots, r$. But V_m can only be V_i, since none of $V_1, \ldots, V_r$ contains another. Thus each V_i equals some unique W_j

and similarly each W_j equals some V_i, whence the uniqueness assertion.

In the expression $V = V_1 \cup \cdots \cup V_r$ of a closed subset V of a noetherian space as the union of irreducible closed subsets none of which contains another, the $V_1, \ldots, V_r$ are called the *components* of V. Each V_i is connected. The space V itself may be reducible but still connected; this will be the case if all the components of V can be arranged in a sequence, possibly with repetitions, so that any two consecutive terms in the sequence have a nonempty intersection.

We define a *morphism* from a subset of k^n into a subset of k^m to be a map from the first set into the second which, in a suitable neighborhood of each point of the first set, can be defined coordinatewise by m rational functions with coefficients in k of the coordinates of k^n, each of these rational functions having denominators which are never zero on the open neighborhood in question. The notion of *isomorphism* from a subset of k^n to a subset of k^m is clear, as is the fact that morphisms are continuous maps, and the fact that the composition of two morphisms is again a morphism.

We define the cartesian product of a subset of k^n with a subset of k^m by identifying $k^n \times k^m$ with k^{n+m}. It happens that if n and m are positive, the topology in $k^n \times k^m$ is strictly stronger than the usual product topology, but the projections on the two factors are morphisms, and the restrictions of these to cross-sections are, as would be wished, isomorphisms.

2. ELEMENTARY PROPERTIES OF LINEAR TRANSFORMATIONS

The material in this section is all widely known, but not quite so widely as to justify its omission.

THEOREM 2: *Let V be a finite dimensional vector space over the algebraically closed field k and let $\{T_\alpha\}_{\alpha \in A}$ be a set of commuting linear transformations on V. Then V is the direct sum of subspaces W which are such that $T_\alpha(W) \subset W$ for each $\alpha \in A$ and the restriction of T_α to W has only one characteristic value. Each such* *subspace W has a basis with respect to which each T_α has a matrix in supertriangular form with equal diagonal elements.*

Let I be the identity linear transformation. For any linear transformation T on V we have

$$(T - c_1 I)^{\nu_1} \cdots (T - c_r I)^{\nu_r} = 0$$

for suitable distinct $c_1, \ldots, c_r \in k$ and suitable positive integers $\nu_1, \ldots, \nu_r$. If x is an indeterminate over k, the polynomials $(x - c_1)^{\nu_1} \cdots (x - c_r)^{\nu_r}(x - c_i)^{-\nu_i}$, $i = 1, \ldots, r$, have no common factor, so there exist $f_1(x), \ldots, f_r(x) \in k[x]$ such that

$$\sum_{i=1}^{r} f_i(x)(x - c_1)^{\nu_1} \cdots (x - c_r)^{\nu_r}(x - c_i)^{-\nu_i} = 1.$$

Therefore

$$I = f_1(T)(T - c_2 I)^{\nu_2} \cdots (T - c_r I)^{\nu_r} + \cdots$$
$$+ f_r(T)(T - c_1 I)^{\nu_1} \cdots (T - c_{r-1} I)^{\nu_{r-1}}.$$

If $T \in \{T_\alpha\}_{\alpha \in A}$, then each of the subspaces of V given by $f_1(T)(T - c_2 I)^{\nu_2} \cdots (T - c_r I)^{\nu_r}(V), \ldots, f_r(T)(T - c_1 I)^{\nu_1} \cdots (T - c_{r-1} I)^{\nu_{r-1}}(V)$ is mapped into itself by any T_α, since T and T_α commute, and V is the sum of these subspaces. These subspaces are also annulled by $(T - c_1 I)^{\nu_1}, \ldots, (T - c_r I)^{\nu_r}$, respectively, and it readily follows from this that V is their direct sum. Therefore V is the direct sum of subspaces each of which is mapped into itself by each T_α and such that the restriction of T to each of these subspaces has only one characteristic value. Applying this fact to all T's in $\{T_\alpha\}_{\alpha \in A}$, we get V to be the direct sum of subspaces W as described. It remains only to show that each W has a basis such as described, and for this it suffices to show that each W has a basis with respect to which each T_α has a matrix in supertriangular form, that is, with only zeros below the main diagonal. It suffices to show that V has a basis with respect to which each T_α has a matrix in supertriangular form. This amounts to finding a sequence of subspaces of V, say

$V = V_0 \supset V_1 \supset V_2 \supset \cdots \supset V_{\dim V} = \{0\}$ such that $T_\alpha(V_i) \subset V_i$ for each $\alpha \in A$ and each i and such that $\dim V_i/V_{i+1} = 1$ for $i = 0, \ldots, \dim V - 1$; for if we choose v_i in $V_{i-1} - V_i$, then $v_1, \ldots, v_{\dim V}$ will be an appropriate basis of V. To prove that we have such a sequence of subspaces of V, it suffices to prove by induction on $\dim V$ that if $\dim V > 1$ then there exists a subspace $U \subset V$ such that $\{0\} \neq U \neq V$ and such that $T_\alpha(U) \subset U$ for all $\alpha \in A$, for if we apply the induction hypothesis to the vector spaces U and V/U and the linear transformations induced on them by $\{T_\alpha\}_{\alpha \in A}$, we get the desired chain $V = V_0 \supset V_1 \supset \cdots \supset \{0\}$. To prove the existence of a subspace U as desired in case $\dim V > 1$ is easy: if each T_α, $\alpha \in A$, is a scalar linear transformation, we can take U to be any subspace between $\{0\}$ and V, while if there exists an $\alpha \in A$ such that T_α is not scalar, we can take U to be the null space of $T_\alpha - cI$, where c is one of the characteristic values of T_α. This completes the proof.

Let T be a linear transformation on a vector space V that is finite dimensional over the field k. We call T *semisimple* if it annuls a nonzero polynomial in one variable with coefficients in k that has no multiple root. We call T *unipotent* if $T - I$ is nilpotent. Clearly I is the only linear transformation that is both semisimple and unipotent.

THEOREM 3: *Let T be a nonsingular linear transformation on the finite dimensional vector space V over the algebraically closed field k. Then there exist unique linear transformations T_s and T_u on V, respectively semisimple and unipotent, such that $T = T_sT_u = T_uT_s$.*

To obtain T_s, T_u (which are called, respectively, the *semisimple part* and the *unipotent part* of T), we simply apply the previous theorem to the set $\{T\}$ consisting of T alone and define T_s to be the linear transformation on V whose restriction to each W is scalar multiplication by the characteristic value of the restriction of T to W, a linear transformation that is clearly semisimple and commutes with T, and we define T_u to be $TT_s^{-1} = T_s^{-1}T$, which is clearly unipotent since its only characteristic value is 1. It remains only to show the uniqueness of T_s and T_u. To do this, go back to the proof of the preceding theorem and note that each of the spaces W there is the image of V by a polynomial with coefficients in k in the various linear transformations $\{T_\alpha\}_{\alpha\in A}$, so that in the present case each W is the image of V by a polynomial in the single linear transformation T, which implies that the T_s we have constructed commutes with any linear transformation on V that commutes with T. Thus also T_u commutes with any linear transformation on V that commutes with T. If now T_s', T_u' are linear transformations on V, respectively semisimple and unipotent, such that $T = T_s'T_u' = T_u'T_s'$, then we apply the preceding theorem to the commuting set of linear transformations $\{T_s, T_u, T_s', T_u'\}$ and note that the matrices of T_s and T_s' with respect to the appropriate basis of V will be diagonal, while the matrices of T_u and T_u' will be supertriangular with all diagonal elements equal to 1. Thus the matrix of $T_sT_s'^{-1} = T_u'T_u^{-1}$ will be on the one hand diagonal and on the other hand supertriangular with all diagonal elements equal to 1, hence will be the identity matrix I. Thus $T_s' = T_s$, $T_u' = T_u$.

3. LINEAR ALGEBRAIC GROUPS

The set of all $n \times n$ matrices with entries in k, which can be identified in an obvious way with the set k^{n^2}, inherits the Zariski topology of the latter, independent of the order in which we take the n^2 matrix entries. The subset of all nonsingular $n \times n$ matrices with entries in k, which consists of those matrices with nonzero determinant, is an open subset, and these constitute the elements of the group GL(n), the group operation being ordinary matrix multiplication. By an *algebraic group of $n \times n$ matrices* we mean a closed subgroup of GL(n). The notion of *homomorphism* for algebraic groups of matrices is quite clear, a homomorphism being a morphism which is at the same time a homomorphism of abstract groups. The notion of *isomorphism* is equally clear. By a *linear algebraic group* we mean an isomorphism class of algebraic groups of matrices. Thus if V is a vector space of dimension n over

k, the group $\mathrm{GL}(V)$ of all nonsingular linear transformations on V can be given a well-defined structure of linear algebraic group, since $\mathrm{GL}(V)$ can be identified with $\mathrm{GL}(n)$ by choosing a specific basis of V, and if we alter the basis we remain within the same isomorphism class. There are obvious notions of *morphism*, *homomorphism*, and *isomorphism* for linear algebraic groups.

Note that an algebraic group of $n \times n$ matrices is isomorphic to an algebraic group of $(n+1)\times(n+1)$ matrices that is a closed subset of $k^{(n+1)^2}$ by means of the map

$$g \mapsto \begin{pmatrix} g & 0 \\ 0 & (\det g)^{-1} \end{pmatrix}.$$

Note also that the cartesian product of algebraic groups of $n \times n$ and $m \times m$ matrices can be considered to be an algebraic group of $(n+m)\times(n+m)$ matrices in a natural way.

Examples of algebraic subgroups (that is, closed subgroups) of $\mathrm{GL}(n)$ that come to mind immediately are the subgroup of all $n \times n$ matrices of determinant 1, the subgroup of all orthogonal matrices, the subgroup of all orthogonal matrices of determinant 1, the subgroup of all nonsingular diagonal matrices, and the subgroup of all nonsingular matrices in supertriangular form. The algebraic group of 2×2 matrices

$$\left\{ \begin{pmatrix} 1 & a \\ 0 & 1 \end{pmatrix} : a \in k \right\}$$

is isomorphic (as a set) to k, with the group operation corresponding to ordinary addition on k.

If G is a linear algebraic group, then the map $G \times G \to G$ given by $(g_1, g_2) \mapsto g_1 g_2^{-1}$ is clearly a morphism, as is the map $G \times G \to G$ given by $(g_1, g_2) \mapsto g_1 g_2$. The map $G \to G$ given by $g \mapsto g^{-1}$ is an isomorphism, in the sense of isomorphism classes of subsets of finite cartesian products of k with itself, as are the maps $G \to G$ given by $g \mapsto ag$ and $g \mapsto ga$ for any fixed $a \in G$.

THEOREM 4: *A linear algebraic group G has a unique component G_0 containing the identity element e, and G_0 is a closed normal* *subgroup of G of finite index. The various components of G are the (left or right) cosets of G_0 in G, and these are disjoint.*

Each component of G is irreducible and not contained in any of the other components, therefore not contained in their union. Thus each component of G contains at least one point not lying on any other component. But G is homogeneous, in the sense that left or right translation by any fixed element of G is an isomorphism of sets and these translations permute the points of G transitively. From this it follows that *any* point of G is contained in only one component of G and that the various components of G, finite in number, are the (left or right) translates of the component G_0 that contains e. It remains only to show that G_0 is a subgroup of G. To do this, it suffices to consider the morphism of sets $G \times G \to G$ given by $(g_1, g_2) \mapsto g_1 g_2^{-1}$ and note that the induced morphism $G_0 \times G_0 \to G$ sends (e, e) into e, so that by continuity the image of $G_0 \times G_0$ is contained in the topological connected component G_0 of G.

The simple topological arguments used above can be applied, with equal ease, to a number of other questions. For example, let us prove that if G is a linear algebraic group and H a subgroup of G, then the closure of H is also a subgroup of G. Using the symbol * to denote closure, all we have to do is note that since the morphism $G \times G \to G$ given by $(g_1, g_2) \mapsto g_1 g_2^{-1}$ sends $H \times H$ into H, because H is a subgroup of G, this morphism therefore sends $H^* \times H^*$ into H^*, by continuity, proving that H^* is a subgroup of G. If H is a connected subgroup of G, then so is H^*. If H is a commutative subgroup of G, then so is H^*, for the relation $h_1 h_2 h_1^{-1} h_2^{-1} = e$ which holds for $h_1, h_2 \in H$ must also hold for all $h_1, h_2 \in H^*$. If $H_1 \subset H_2$ are subgroups of G with H_1 normal in H_2, then H_1^* is normal in H_2^*; this fact comes from consideration of the morphism $G \times G \to G$ given by $(g_1, g_2) \mapsto g_2 g_1 g_2^{-1}$, which maps $H_1 \times H_2$ into H_1 and so by continuity maps $H_1^* \times H_2^*$ into H_1^*. Let us now prove that if S is a connected subset of G that contains e, then the subgroup of G generated by S is connected. To do this, it suffices to assume that S contains the inverse of each of its elements, for otherwise we just replace S by the connected set which is the union of S with the set of all inverses of its elements. The continuity of group multiplication shows that the sets $S^2 = \{s_1 s_2 : s_1, s_2 \in S\}$, $S^3 = \{s_1 s_2 s_3 : s_1, s_2, s_3 \in S\}$, ... are all connected, and hence so is $S \cup S^2 \cup S^3 \cup \cdots$, which is just the group generated by S. A special consequence of what we have just shown is that the commutator subgroup of a connected linear algebraic group is also connected; for this subgroup of G is generated by all commutators $g_1 g_2 g_1^{-1} g_2^{-1}$, with $g_1, g_2 \in G$, and these commutators are a connected set containing e. This fact will be used later.

THEOREM 5: *Let G be a closed subgroup of the group of all diagonal matrices in* GL(n). *Then there exists a subset $S \subset \mathbf{Z}^n$ such that G consists of all diagonal matrices with nonzero entries $x_1, \ldots, x_n$ in k such that $x_1^{\nu_1} x_2^{\nu_2} \cdots x_n^{\nu_n} = 1$ for all $(\nu_1, \ldots, \nu_n) \in S$.*

The group of all diagonal matrices in GL(n) may be identified in an obvious way with the group $(\mathrm{GL}(1))^n$. For $i = 1, \ldots, n$, let X_i denote the ith coordinate function on this cartesian product. Suppose that the nonzero polynomial $F(X) = \sum c_{\nu_1 \cdots \nu_n} X_1^{\nu_1} \cdots X_n^{\nu_n}$ vanishes on G, where each $c_{\nu_1 \cdots \nu_n}$ is a nonzero element of k and $(\nu_1, \ldots, \nu_n)$ ranges over a finite set of n-tuples of nonnegative integers. Then for each $a = (a_1, \ldots, a_n) \in G$, the polynomial $F(aX) = F(a_1 X_1, \ldots, a_n X_n) = \sum c_{\nu_1 \cdots \nu_n} a_1^{\nu_1} \cdots a_n^{\nu_n} X_1^{\nu_1} \cdots X_n^{\nu_n}$ also vanishes on G. Unless all $a_1^{\nu_1} \cdots a_n^{\nu_n}$ are equal, a suitable linear combination with coefficients in k of $F(X)$ and $F(aX)$ will be a nonzero polynomial vanishing on G that contains only terms $X_1^{\nu_1} \cdots X_n^{\nu_n}$ appearing in $F(X)$ and that will have strictly fewer nonzero terms than $F(X)$. Then $F(X)$ will be the sum of two other polynomials vanishing on G, each of which contains only terms $X_1^{\nu_1} \cdots X_n^{\nu_n}$ appearing in $F(X)$ and each of which has strictly fewer nonzero terms than $F(X)$. Repeating this process enough times, using all $(a_1, \ldots, a_n) \in G$, we get $F(X)$ to be a linear combination with coefficients in k of polynomials $F_1(X), \ldots, F_r(X)$ that vanish on G, each of which has only terms that appear in $F(X)$ and each of which is minimal in the obvious sense of not being the sum of such polynomials with fewer terms. If $F(X)$ itself is minimal then clearly it has at least two terms, say $X_1^{\nu_1} \cdots X_n^{\nu_n}$ and $X_1^{\mu_1} \cdots X_n^{\mu_n}$, and we have $a_1^{\nu_1} \cdots a_n^{\nu_n} = a_1^{\mu_1} \cdots a_n^{\mu_n}$ for all $(a_1, \ldots, a_n) \in G$; thus $X_1^{\nu_1} \cdots X_n^{\nu_n} - X_1^{\mu_1} \cdots X_n^{\mu_n}$ vanishes on G and $F(X)$ itself is the product of this latter polynomial by an element of k. The totality of minimal polynomials in $X_1, \ldots, X_n$ that vanish on G gives us the desired monomial equations.

COROLLARY: *A proper closed subgroup of the group of all diagonal matrices in* GL(n) *is isomorphic to a direct product* $(\mathrm{GL}(1))^m \times H$, *where* $0 \leqslant m < n$ *and* H *is the direct product of* $n - m$ *cyclic groups of finite order not divisible by the characteristic of* k.

The set S of all n-tuples $(\nu_1, \ldots, \nu_n) \in \mathbf{Z}^n$ such that $x_1^{\nu_1} \cdots x_n^{\nu_n} = 1$ for each $(x_1, \ldots, x_n)$ in our proper closed subgroup G of $(\mathrm{GL}(1))^n$ is an additive subgroup of $\mathbf{Z}^n$. The structure theory of finitely generated abelian groups implies the existence of a free set of generators $\{(a_{1i}, \ldots, a_{ni})\}_{i=1,\ldots,n}$ for the group $\mathbf{Z}^n$ and integers $d_1, \ldots, d_n \geqslant 0$ such that the group S is generated by $\{(d_i a_{1i}, \ldots, d_i a_{ni})\}_{i=1,\ldots,n}$. The map $(x_1, \ldots, x_n) \mapsto (x_1^{a_{11}} \cdots x_n^{a_{n1}}, \ldots, x_1^{a_{1n}} \cdots x_n^{a_{nn}})$ is an automorphism of the group $(\mathrm{GL}(1))^n$ which sends G onto the subgroup given by the equations $\{X_i^{d_i} = 1\}_{i=1,\ldots,n}$. In the case of characteristic $p \neq 0$, if $p \mid d_i$ the equation $x_i^{d_i} = 1$ is equivalent to the equation $x_i^{d_i/p} = 1$, so that we may assume that p does not divide any nonzero d_i. Since G is a proper subgroup, not all d_i's are 0, and the result is now clear.

THEOREM 6: *The smallest closed subgroup of* GL(n) *which contains a given unipotent matrix* $I + a$ *is, in characteristic zero,* $\{\exp tb : t \in k\}$, *where* exp *denotes the usual exponential series and* $b = \log(I + a) = a - a^2/2 + a^3/3 - \cdots$; *if* $a \neq 0$ *this is a group isomorphic to* k *with its additive group structure under the correspondence* $\exp tb \leftrightarrow t$. *In characteristic* $p \neq 0$, *this smallest closed subgroup containing the unipotent matrix* $I + a$ *is cyclic of order a power of* p.

Since $I + a$ is unipotent, a is a nilpotent $n \times n$ matrix. If the field characteristic is $p \neq 0$, we have $a^{p^\nu} = 0$ for ν sufficiently large, implying $(I + a)^{p^\nu} = I$ and the truth of the theorem. So let us assume that we are in the case of characteristic zero. Formal computation shows that log gives an isomorphism between the closed set of all unipotent matrices and that of all nilpotent matrices, the inverse isomorphism being given by exp. Thus the set $\{\exp tb : t \in k\}$ is closed and contains the point $\exp b = I + a$. If α and β are commuting $n \times n$ nilpotent matrices, then $\exp \alpha \cdot \exp \beta = \exp(\alpha + \beta)$; in particular, $\exp t_1 b \cdot \exp t_2 b = \exp(t_1 + t_2)b$. Thus if $a \neq 0$, the set $\{\exp tb : t \in k\}$ is a closed subgroup of GL(n) that is isomorphic to k with its additive group structure. Since the only closed subsets of k are k itself and its finite subsets, the only proper closed subgroup of k in characteristic zero is $\{0\}$, so we are done.

LEMMA: *Let G be a connected closed subgroup of* $(\mathrm{GL}(1))^n \times k$, *with the additive group structure on k. Then either* $G \subset (\mathrm{GL}(1))^n \times \{0\}$ *or* $G \supset \{1\} \times \cdots \times \{1\} \times k$.

We prove this by induction on n, the case $n = 0$ reducing to the known circumstance that the only connected closed subsets of k are k itself and its single points. We therefore assume that $n > 0$ and choose coordinate functions $X_1, \ldots, X_n, Y$ on the factors $\mathrm{GL}(1), \ldots, \mathrm{GL}(1), k$ so that the group law is given by $(x_1, \ldots, x_n, y)(x'_1, \ldots, x'_n, y') = (x_1x'_1, \ldots, x_nx'_n, y + y')$. If $G \neq (\mathrm{GL}(1))^n \times k$, then there is a nonzero polynomial function $F(X_1, \ldots, X_n, Y) \in k[X_1, \ldots, X_n, Y]$ that vanishes on G. The set of all functions in $k[X_1, \ldots, X_n, Y]$ that vanish on G and have at most a prescribed total degree is a finite dimensional vector space over k. G operates on this vector space as a group of linear transformations, by means of translation of functions: if $\gamma = (x_1, \ldots, x_n, y) \in G$ and $f \in k[X_1, \ldots, X_n, Y]$ then $f_\gamma \in k[X_1, \ldots, X_n, Y]$ is given by $f_\gamma(X_1, \ldots, X_n, Y) = f(x_1X_1, \ldots, x_nX_n, Y + y)$. Since G is commutative, by Theorem 2 a basis for this vector space of polynomial functions vanishing on G can be chosen with respect to which each linear transformation induced by an element of G has a matrix in supertriangular form, and one of these basis elements will be a characteristic vector for each translation. That is, there is a nonzero $\Phi(X_1, \ldots, X_n, Y) \in k[X_1, \ldots, X_n, Y]$ that vanishes on G and has the property that for each $(x_1, \ldots, x_n, y) \in G$ we have $\Phi(x_1X_1, \ldots, x_nX_n, Y+y) = C(x_1, \ldots, x_n, y)\Phi(X_1, \ldots, X_n, Y)$, with $C(x_1, \ldots, x_n, y) \in k$. If we write $\Phi(X_1, \ldots, X_n, Y) = \sum_{i_1, \ldots, i_n \geqslant 0} X_1^{i_1} \cdots X_n^{i_n} F_{i_1 \cdots i_n}(Y)$, with the various $(i_1, \ldots, i_n)$ distinct and each $F_{i_1 \cdots i_n}(Y) \in k[Y]$, we get that for each $(i_1, \ldots, i_n)$ and each $(x_1, \ldots, x_n, y) \in G$ we have $F_{i_1 \cdots i_n}(Y+y)$ a multiple of $F_{i_1 \cdots i_n}(Y)$, and thus $F_{i_1 \cdots i_n}(Y+y) = F_{i_1 \cdots i_n}(Y)$. If $G \not\subset (\mathrm{GL}(1))^n \times \{0\}$, then since G is connected it contains points $(x_1, \ldots, x_n, y)$ with an infinite number of distinct y's, proving that each $F_{i_1 \cdots i_n}(Y) \in k$. Furthermore Φ has more than one term, and if the distinct terms $X_1^{i_1} \cdots X_n^{i_n}, X_1^{j_1} \cdots X_n^{j_n}$ actually occur in Φ, then for each $(x_1, \ldots, x_n, y) \in G$ we have $x_1^{i_1} \cdots x_n^{i_n} = x_1^{j_1} \cdots x_n^{j_n}$. Thus there is a proper closed subgroup H of $(\mathrm{GL}(1))^n$ such that $G \subset H \times k$. Since G is connected, we can take H connected. Then H is isomorphic to a product $\mathrm{GL}(1) \times \cdots \times \mathrm{GL}(1)$ with fewer than n factors, and the induction procedure goes through.

The following theorem tells us, among other things, that the splitting $g = g_s g_u$ of an element g of an algebraic group of matrices into its semisimple and unipotent parts is independent of the matrix representation of the group, so that it is permissible to speak of the semisimple and unipotent parts of the elements of a linear algebraic group. Also, for any homomorphism $\tau : G \to G'$ of linear algebraic groups, for each $g \in G$ we have $(\tau(g))_s = \tau(g_s)$, $(\tau(g))_u = \tau(g_u)$, so that the semisimple-unipotent decomposition of $\tau(g)$ is $\tau(g) = \tau(g_s)\tau(g_u)$.

THEOREM 7: *If G is a closed subgroup of* GL(n) *and $g \in G$, then also $g_s, g_u \in G$. Furthermore, for any homomorphism $\tau : G \to \mathrm{GL}(m)$, $\tau(g_s)$ and $\tau(g_u)$ are, respectively, semisimple and unipotent.*

We may assume that G is the smallest closed subgroup of $\mathrm{GL}(n)$ that contains g. Our first task is to prove that $g_s, g_u \in G$. If the field characteristic is $p \neq 0$, there is a positive integer ν such that $g_u^{p^\nu} = I$. Then $g^{p^\nu} = g_s^{p^\nu}$ is semisimple. Now G contains the smallest closed subgroup of $\mathrm{GL}(n)$ containing the semisimple element $g_s^{p^\nu}$, which is commutative and consists entirely of elements commuting with g. The smallest closed subgroup of $\mathrm{GL}(n)$ containing any given semisimple element is isomorphic to a closed subgroup of $(\mathrm{GL}(1))^n$, and on any such group the endormorphism $\gamma \mapsto \gamma^{p^\nu}$ is surjective. Hence G contains a semisimple element g_1 which commutes with g and is such that $g_1^{p^\nu} = g^{p^\nu}$. Thus $(gg_1^{-1})^{p^\nu} = I$, implying that gg_1^{-1} is unipotent. Therefore $g = g_1 \cdot gg_1^{-1}$ with g_1 and gg_1^{-1} commuting and, respectively, semisimple and unipotent. Therefore $g_s = g_1$ and $g_u = gg_1^{-1}$ and they are both in G. If k has characteristic zero, the proof goes differently. In this case we start with the permissible assumption that g is neither semisimple nor unipotent. Let S, U be the smallest closed subgroups of $\mathrm{GL}(n)$ that contain g_s and g_u, respectively. S is isomorphic to a closed subgroup of $(\mathrm{GL}(1))^n$, while U is isomorphic to k with its additive group structure. Since g_s and g_u commute, each element of S commutes with g_u and each element of U commutes with each element of S, so that the subgroup SU of $\mathrm{GL}(n)$ is commutative. Consider now the morphism $S \times U \to \mathrm{GL}(n)$ given by $(s,u) \mapsto su$; this is actually a homomorphism, since SU is commutative. The inverse image of G is the closed subgroup Γ of $S \times U$ consisting of all $(s,u) \in S \times U$ such that $su \in G$. We have $(g_s, g_u) \in \Gamma$, so if $N = [\Gamma : \Gamma_0]$, where, as before, Γ_0 is the component of the identity of Γ, then $(g_s^N, g_u^N) \in \Gamma_0$. Now $g_u \neq I$, so $g_u^N \neq I$, so it follows from the lemma that $\Gamma_0 \supset I \times U$, which in turn implies that $G \supset U$. Thus $g_u, g_s \in G$, which completes the proof of the first statement. Before proceeding with the proof of the second statement we remark that on any closed commutative subgroup of $\mathrm{GL}(n)$, the maps $g \mapsto g_s$, $g \mapsto g_u$ are homomorphisms; it suffices to prove this for the map $g \mapsto g_s$, which is easily done by putting the group G into the supertriangular form of Theorem 2, in which case g_s is obtained from g by taking the matrix representation of the latter and replacing all the off-diagonal elements by 0, a map that is clearly both a morphism and a group-theoretic homomorphism. The homomorphism τ of the theorem maps G into the smallest algebraic subgroup of $\mathrm{GL}(m)$ that contains $\tau(g)$. Our preceding remarks show that to complete the proof of the theorem it suffices to show that if g is semisimple (or unipotent) and $\tau(g)$ is unipotent (or semisimple), then τ is trivial on G. If g is semisimple then G has the structure described in the corollary to Theorem 5, which implies that the elements of G of finite order not divisible by the field characteristic are dense in G. These elements of G have images under τ that are both unipotent and of finite order not divisible by the characteristic of k, hence must all be the identity element of $\mathrm{GL}(m)$. Thus τ is trivial on G, hence on g. In the case where g is unipotent and $\tau(g)$ semisimple, τ maps G into a group isomorphic to a closed subgroup of $(\mathrm{GL}(1))^m$, so to prove that τ is trivial we may as well assume that τ maps G into $\mathrm{GL}(1)$. If the field characteristic is $p \neq 0$ then g has order a power of p, say $g^{p^r} = I$, so that $(\tau(g))^{p^r} = 1$ and therefore $\tau(g) = 1$. In the case of characteristic zero, the coordinate function on $\mathrm{GL}(1)$ induces, via τ and via the isomorphism of G with the additive group of k given by Theorem 6, assuming $g \neq I$, a function on k that in a neighborhood of each point equals the quotient of two polynomial functions neither of which vanishes anywhere on the neighborhood. This gives a rational function on k which is finite and nonzero everywhere, hence a nonzero constant, so that here too τ is trivial.

THEOREM 8: *If G is a commutative linear algebraic group, then the set G_s of all semisimple elements of G and the set G_u of all unipotent elements of G are closed subgroups of G and the map $G_s \times G_u \to G$ given by $(s, u) \mapsto su$ is an isomorphism.*

As remarked in the course of the proof of the preceding theorem, if G is taken to be a closed subgroup of $\mathrm{GL}(n)$ that is in the supertriangular form of Theorem 2, then the map $g \mapsto g_s$ on G is obtained by replacing any matrix by the matrix with the same diagonal elements and with zero elsewhere. The map $G \to G$ given by $g \mapsto g_s$ is therefore a homomorphism, as is the map $g \mapsto g_u$. For

G in our special supertriangular form, G_s is the intersection of G with the closed subgroup of all diagonal matrices of $\mathrm{GL}(n)$, while G_u is the closed subset of unipotent elements of G. Thus G_s and G_u are closed subgroups of G. The map $G_s \times G_u \to G$ given by $(s,u) \mapsto su$ is clearly a surjective morphism which is a group-theoretic isomorphism. Since the projections $G \to G_s$, $G \to G_u$ are morphisms, the given map $G_s \times G_u \to G$ is an isomorphism.

We conclude this article with a proof of the well-known Lie-Kolchin theorem on solvable linear algebraic groups.

THEOREM 9: *Let V be a finite dimensional vector space over k, G a connected solvable subgroup of* $\mathrm{GL}(V)$. *Then there is a basis of V with respect to which the matrices of the elements of G are all in supertriangular form.*

The solvability of G is here understood in the usual sense, one formulation of which is that the descending chain of subgroups of G that starts with G and takes as every other group in the chain the commutator subgroup of its predecessor terminates in $\{e\}$. Subgroups and factor groups of solvable groups are solvable. Since it was shown immediately before the statement of Theorem 5 that the commutator subgroup of a connected linear algebraic group is connected, and since the chain of commutator subgroups of G is invariant under all automorphisms of G, we can state that our present group G is either commutative or that its commutator subgroup contains a connected commutative subgroup $H \neq \{e\}$ that is a normal subgroup of G. Our present theorem is trivial if $\dim V = 1$, so that we may assume $\dim V > 1$ and proceed to a proof by induction on $\dim V$. The theorem is true if G is commutative, even without the connectedness assumption, by Theorem 2, so that G may be assumed to be noncommutative. If V contains a nontrivial G-invariant subspace W, that is, a subspace W other than $\{0\}$ or V such that $g(W) \subset W$ for all $g \in G$, then each element of G induces a linear transformation on each of the vector spaces W and V/W, and thus we obtain homomorphisms from G into $\mathrm{GL}(W)$ and $\mathrm{GL}(V/W)$. Each image of G will clearly be both solvable and connected, so by our induction assumption there will be bases of W and of V/W with respect to which the actions of G are given by matrices in supertriangular form. If we now choose a basis of V consisting of this special basis of W and elements of V mapping onto the appropriate basis of V/W, then we obtain a basis of V with respect to which each element of G has a matrix in supertriangular form, thus proving the theorem. We may therefore assume that V has no nontrivial G-invariant subspaces. Now the normal subgroup H of G is commutative, so that Theorem 2 may be applied to it to get a nonzero $v \in V$ that is a simultaneous characteristic vector for all elements of H, meaning that there is a scalar-valued function χ on H such that $h(v) = \chi(h)v$ for all $h \in H$. If $\gamma \in G$ and $h \in H$, then $h(\gamma(v)) = \gamma((\gamma^{-1}h\gamma)(v)) = \gamma(\chi(\gamma^{-1}h\gamma)v) = \chi(\gamma^{-1}h\gamma)\gamma(v)$, so that $\gamma(v)$ is also a simultaneous characteristic vector for all elements of H. Therefore the subspace of V spanned by all simultaneous characteristic vectors for all elements of H is a G-invariant subspace. By assumption, this subspace is V itself. Thus V has a basis with respect to which each element of H is in diagonal form. Conjugation by elements of G gives automorphisms of H, hence automorphims of the closure H^* of H in $\mathrm{GL}(V)$. But H^* is isomorphic to a closed subgroup of $(\mathrm{GL}(1))^{\dim V}$, so the corollary of Theorem 5 implies that its points of any given finite order N are finite in number and that the set of all such points for all N is dense in H^*. Since conjugation by elements of G sends points of order N into points of order N, the connectedness of G shows that this conjugation leaves each point of H^* of order N fixed, therefore each point of any finite order fixed, so that G commutes with each element of H^*. Therefore H is in the center of G. If the nonzero $v \in V$ is a simultaneous characteristic vector for all elements of H, with $h(v) = \chi(h)v$ for $h \in H$, our earlier computation now shows that for each $\gamma \in G$ we have $h(\gamma(v)) = \chi(h)\gamma(v)$, so that the subspace of V consisting of all v' such that $h(v') = \chi(h)v'$ for all $h \in H$ is a G-invariant subspace of V, therefore all of V. Therefore H consists entirely of scalar linear transformations. Since H is a subgroup of the commutator subgroup of G, each element of H has determinant 1. Thus H is a finite group, necessarily $\{e\}$ since H is connected, a contradiction which completes the proof of the theorem.

It is worth remarking that the converse of the theorem is clearly true, that is, that a group of matrices in supertriangular form is solvable. The easiest way to see this is to exhibit a normal chain for the full supertriangular group whose successive factor groups are all commutative, as for example, for 4×4 matrices,

$$\left\{\begin{bmatrix} x & x & x & x \\ 0 & x & x & x \\ 0 & 0 & x & x \\ 0 & 0 & 0 & x \end{bmatrix}\right\} \supset \left\{\begin{bmatrix} 1 & x & x & x \\ 0 & 1 & x & x \\ 0 & 0 & 1 & x \\ 0 & 0 & 0 & 1 \end{bmatrix}\right\} \supset \left\{\begin{bmatrix} 1 & 0 & x & x \\ 0 & 1 & 0 & x \\ 0 & 0 & 1 & 0 \\ 0 & 0 & 0 & 1 \end{bmatrix}\right\}$$

$$\supset \left\{\begin{bmatrix} 1 & 0 & 0 & x \\ 0 & 1 & 0 & 0 \\ 0 & 0 & 1 & 0 \\ 0 & 0 & 0 & 1 \end{bmatrix}\right\} \supset \left\{\begin{bmatrix} 1 & 0 & 0 & 0 \\ 0 & 1 & 0 & 0 \\ 0 & 0 & 1 & 0 \\ 0 & 0 & 0 & 1 \end{bmatrix}\right\}.$$

We know that the conclusion of Theorem 9 holds for G commutative, whether or not G is connected. But it may fail if G is solvable but not connected. For example, let $G \subset \mathrm{GL}(n)$ be a finite noncommutative group of order not divisible by the characteristic of k. Such a G cannot be put into supertriangular form, for otherwise each commutator of G would be both unipotent and of finite order not divisible by the field characteristic, hence semisimple, hence equal to the identity $\{e\}$, which is impossible.

We note that the subgroup G of $\mathrm{GL}(V)$ was not assumed to be closed. It is in fact an easy exercise to show that if G is solvable, then so is its closure G^* in $\mathrm{GL}(V)$ and that indeed G^* has a normal chain of closed subgroups terminating with $\{e\}$ with each successive factor group commutative.

THE CONNECTEDNESS THEOREM AND THE CONCEPT OF MULTIPLICITY

B. L. van der Waerden

In 1926, as a basis of a theory of intersection multiplicities, I developed the concept of the multiplicity of a specialization and under certain assumptions showed the uniqueness of the multiplicity [1]. The line of thought was as follows:

Let a "normal problem" be given by equations

$$G_j(\xi, \eta) = 0, \tag{A}$$

where the G_j are homogeneous in the unknowns $\eta_0, \ldots, \eta_n$ and the ξ are indeterminates. Assume that problem (A) has only a finite number of solutions $\eta^{(1)}, \ldots, \eta^{(h)}$ for indeterminate ξ. Every specialization $\xi \to x$ can be extended to a specialization:

$$(\xi, \eta^{(1)}, \ldots, \eta^{(h)}) \to (x, y^{(1)}, \ldots, y^{(h)}).$$

Translated from B. L. van der Waerden's "Zur algebraischen Geometrie 20: Der Zusammenhangsatz und der Multiplizitätsbegriff," *Mathematische Annalen*, vol. 193 (1971), pp. 89–108, by A. Seidenberg.

The $y^{(\nu)}$ are solutions of the specialized problem

$$G_j(x, y) = 0. \tag{B}$$

The uniqueness of the specialized solutions $y^{(1)}, \ldots, y^{(h)}$ was shown in [1] under the assumption that the specialized problem (B) has only a finite number of solutions for given x. If the solution y of problem (B) occurs exactly μ times amongst the $y^{(\nu)}$, one says that it has *multiplicity* μ in the specialization $\xi \rightarrow x$.

Weil [2] established the uniqueness of the multiplicity μ of a solution y in more general circumstances. Namely, Weil does not require that problem (B) should have only a finite number of solutions, but only that the solution y should be isolated. Northcott [3] and Leung [4] have weakened the assumptions still further.

Weil's proof of the uniqueness of the multiplicity of an isolated solution y is rather complicated: it depends on the analytic theory of local rings. Northcott and Leung also use this theory. In the following a proof will be given using classical methods. The proof depends on a special case of the connectedness theorem.

To explain the *Connectedness Theorem* we need some definitions.

A variety V in projective space P^n is called *connected* if it is not the union of two disjoint non-empty subvarieties; here, if V is defined over the field k, the connectedness condition is also to hold over any extension of k. A variety is called *linearly connected* if any pair of points of V can be connected by a chain of rational curves $C_1, \ldots, C_h$ lying on V. From this it obviously follows that V is connected.

A *cycle* Z^d is a formal sum of absolutely irreducible varieties V^d, all of the same dimension d, with integer coefficients e_i. We consider only positive cycles, for which the e_i are all positive or zero. The union of the V^d with $e_i \neq 0$ is called the *support* of the cycle.

Let, then, ξ be a general and x a special point of an irreducible variety U defined over k. To the point ξ let there be associated a connected cycle defined over $k(\xi)$. Assume that U is analytically irreducible at x, i.e., that the completion of the local ring [at x] is free of zero-divisors. To the specializations $\xi \to x$ belong specializations Z_x of the cycle Z_ξ. Then the connectedness theorem says that the union of the supports of all the Z_x is connected.

The connectedness theorem was originally formulated by Zariski [5] and proved in the context of his theory of holomorphic functions on an algebraic variety. Chow [6] generalized the theorem and gave a simpler proof.

For the application that we here have in view, a special case of the connectedness theorem is of particular significance. The cycle Z_ξ will be a single point η, rationally dependent on ξ. In this case one can strengthen the connectedness theorem, as Chow showed, and prove that the union of the specializations y of η belonging to $\xi \to x$ form a *linearly connected* variety.

In the present work it is first assumed that U is a projective space P^m. The theorem that is to be proved here thus reads (§5, Theorem 2):

In a rational mapping of P^m into P^n the image variety V_A of any point A is linearly connected.

In §1 only basic concepts are explained. In §2 and §3 some known results of classical algebraic geometry are proved using classical methods. For the reader acquainted with the subject, §1 to §3 contain nothing new.

In §4 the above theorem is proved for $U = P^1$ and $U = P^2$. In §5 the case $U = P^m$ is reduced to the case $U = P^2$.

In §6 it is shown that the uniqueness of the multiplicity of an isolated solution y in the case $U = P^m$ follows directly from Theorem 2.

Once one has the uniqueness of the multiplicity of a specialization, one can, following A. Weil, define the intersection multiplicities of cycles. This will be looked into more closely in §7.

In §8 the linear connectedness theorem is proved for the case that x is a simple point of an arbitrary variety U. The general case is reduced to the case $U = P^m$ by using a parallel projection.

1. BASIC CONCEPTS

Let k be a fixed ground field and Ω a *universal domain* in the sense of Weil [2], i.e., an algebraically closed extension field of k having infinite degree of transcendency over k. A *correspondence* K between P^m and P^n is the set of point-pairs (x, y) with coordinates in Ω that satisfy a system of [doubly] homogeneous equations

$$G_j(x, y) = 0 \tag{1}$$

with coefficients in k. For the x-space we often place $x_0 = 1$ and introduce non-homogeneous coordinates, but the y-space will always be the full projective space P^n.

If the correspondence K is irreducible over k, then it consists of all specializations (x, y) of a general pair (ξ, η). One calls (x, y) a *specialization* of (ξ, η) over k if all the homogeneous equations $H(\xi, \eta) = 0$ with coefficients in k that hold for (ξ, η) also hold for (x, y). The specializations of ξ form the *object variety* U, those of η, the *image variety* V of the correspondence, and one speaks of a *correspondence* K *between* U *and* V.

The correspondence K is called a *rational mapping* if the coordinate-ratios of η are rational functions of those of ξ. One then has

$$\beta\eta_k = F_k(\xi). \tag{2}$$

Here the F_k are forms of like degree in $\xi_0, \ldots, \xi_n$ and not all zero.

Amongst the homogeneous equations $H(\xi, \eta) = 0$ that are satisfied by the general pair (ξ, η) of the rational mapping (2), there are the following:

$$\eta_j F_k(\xi) - \eta_k F_j(\xi) = 0. \tag{3}$$

Therefore every pair (x, y) must satisfy

$$y_j F_k(x) - y_k F_j(x) = 0 \tag{4}$$

and hence

$$\gamma y_k = F_k(x). \tag{5}$$

If the $F_k(x)$ are not all zero, then the point x is called *regular* for the mapping; it then has a unique image point y, determined by (5). If all the $F_k(x)$ are zero, then x is *singular* for the mapping; it can then have several images. In every case the set of image points y of a given point x form a variety V_x; one calls it the image variety of the point x in the rational mapping given by (2).

In §2 it will be shown how the individual points y of the image variety of V_x can be obtained by using power series. These power series are the most important tool in the proof, in §4 and §5, of the linear connectedness of V_x.

We need some concepts from the theory of algebraic curves. A curve C in P^m can always be birationally transformed into a curve C' in P^n having no singularities (see, for example, my *Einführung in die algebraische Geometrie*, §45). To a point A of C there can correspond several points A' of C'; each of these defines a *branch* z of the curve C at A. One calls A the *center* of the branch z. If C'' is another [non-singular]* birational transform of C, then the points of C'' are in 1-1 correspondence [via the induced birational mapping] with those of C', so the branch concept is independent of the choice of the transform C'.

For each branch z one can select a uniformizing parameter τ and develop the coordinates of a general point of C in power series in τ:

$$X_i(\tau) = a_i + b_i\tau + \cdots. \tag{6}$$

Here the a_i are the coordinates of the center A of the branch z.

We cut the curve C with a hypersurface $F = 0$ of the space P^m and ask how one gets the intersection multiplicity $(C \cdot F)_A$ at the point A. Let the hypersurface $F = 0$ be given by a form

$$F(X) = F(X_0, X_1, \ldots, X_m).$$

If one wishes to be precise, one must distinguish between the *hypersurface* $F = 0$ and the $(m-1)$-dimensional *cycle* F, which is obtained as follows: The form F is factored over the algebraic closure $\bar{k}$ of k into irreducible factors P_i with exponents e_i. Each of

* Square brackets are used to indicate additions made by the editor.

these irreducible factors P_i defines an irreducible hypersurface $P_i = 0$, which for brevity may be denoted P_i. If these hypersurfaces are counted e_i times and added, one obtains the $(m-1)$-dimensional cycle

$$F = \sum e_i P_i.$$

It is convenient to use the same letter F for the form $F(X)$ and for the cycle F.

The *intersection multiplicity* $(z \cdot F)_A$ of a branch z with the cycle F at the point A can be defined by substituting the branch expansion $X_i(\tau)$ into the form F. If

$$F(X_i(\tau)) = c\tau^{\mu} + \cdots \qquad (c \neq 0),$$

then μ is the intersection multiplicity:

$$(z \cdot F)_A = \mu.$$

The intersection multiplicity $(C \cdot F)_A$ of the curve C with the cycle F at the point A is the sum of intersection multiplicities of the various branches z of the curve with F at A:

$$(C \cdot F)_A = \sum (z \cdot F)_A. \tag{7}$$

For the proof of formula (7) see my *Einführung in die algebraische Geometrie*, §20. For present purposes it would suffice to think of formula (7) as the definition of the intersection multiplicity of the curve C with the cycle F.

2. HOW DOES ONE OBTAIN ALL THE IMAGE POINTS OF A POINT x^0 IN A RATIONAL MAP?

Let an irreducible correspondence K be defined by equations (1). Using the substitution

$$z_{ik} = x_i y_k \tag{8}$$

one can map K onto an image variety in a projective space P^N. We denote the image variety also by K; let its dimension be d.

If in particular K is a rational mapping (2), then the *singular pairs* (x, y), for which all the $F_k(x)$ are zero, form a subvariety K' of dimension at most $d-1$. Let (x^0, y^0) be a singular pair, with image point z^0. Through z^0 we pass a hyperplane u containing no component of K'. By intersecting with u the dimensions of K and K' are each reduced by 1. This process is repeated; after $d-1$ steps one obtains a curve C^* on K, which goes through z^0 and meets K' in only a finite number of points. Let C be an absolutely irreducible component through z^0 of such a curve C^*. Let Z be a general point of C. The corresponding pair (X, Y) is regular, because Z is not in K'.

At least one branch z of the curve C has its center at z^0. If τ is a uniformizing parameter for the branch z, then the coordinates of the points X and Y are represented by power series in τ.

In a neighborhood of the point x^0 one can introduce non-homogeneous coordinates, with x^0 as origin. The power series expansions for the coordinates of the points X and Y then appear as:

$$X_i(\tau) = a_i\tau + b_i\tau^2 + \cdots \qquad (i = 1, \ldots, m), \tag{9}$$

$$Y_k(\tau) = c_k + d_k\tau + e_k\tau^2 + \cdots \qquad (k = 1, \ldots, n). \tag{10}$$

Since the pair $X(\tau)$, $Y(\tau)$ is regular, one must have

$$\gamma Y_k(\tau) = F_k(X(\tau)) \quad \text{with} \quad \gamma \neq 0. \tag{11}$$

If the power series $X(\tau)$ are given, one can expand the right-hand side of (11). If the power series so obtained begins with τ^r, one can write

$$F(X(\tau)) = p_k\tau^r + q_k\tau^{r+1} + \cdots, \tag{12}$$

where the p_k are not all zero. One can take $\gamma = \tau^r$ on the left-hand side of (11) to obtain

$$\tau^r Y_k(\tau) = p_k\tau^r + q_k\tau^{r+1} + \cdots$$

or

$$Y_k(\tau) = p_k + q_k\tau + \cdots. \tag{13}$$

Placing $\tau = 0$, one obtains the center y^0 of the branch:

$$y_k^{(0)} = p_k. \tag{14}$$

From this there follows the desired result:

THEOREM 1: *Every image point y^0 of x^0 can be obtained in the following way. One chooses on U a branch with center x^0 so that the general point $X(\tau)$ of the branch is regular for the mapping. One develops the coordinates $X_i(\tau)$ as power series in τ. One substitutes these power series in the polynomial F_k and expands according to* (12). *Then the leading coefficients p_k of the coordinates are the coordinates of an image point y^0.*

In the following we will describe the process briefly as follows: "One approaches the point x^0 along the branch z and obtains y^0 as limit point."

Of course one can apply the same process to regular x^0; but in this case the result is independent of the choice of the branch.

In particular, if U is the whole space P^m, then one can take the coefficients $a_i, b_i, \ldots$ in the power series (9) in a completely arbitrary way, subject only to the condition that the $F_k(X(\tau))$ do not all become zero. One can use this freedom to break off the power series (9) after the τ^r term: the higher degree terms have no influence on y^0. One can therefore take the $X_i(\tau)$ to be polynomials in τ.

3. REDUCTION OF THE INTERSECTION MULTIPLICITY OF PLANE CURVES BY CREMONA TRANSFORMATIONS

In his fundamental works on the resolution of singularities of algebraic curves, Max Noether showed that the intersection multiplicity $(F, G)_A$ of two plane curves F and G at the point A can always be reduced by a suitable Cremona transformation. Noether used the Cremona transformation

$$x_0 = z_1 z_2, \qquad x_1 = z_2 z_0, \qquad x_2 = z_1 z_0. \tag{15}$$

If F has an r-fold point at A and G has an s-fold one there, then the intersection multiplicity, according to Noether, is reduced rs by the transformation.

Instead of the transformation (15) one can also use the somewhat simpler transformation

$$x_1 = z_1, \qquad x_2 = z_1 z_2. \tag{16}$$

The non-homogeneous coordinates (x_1, x_2) are chosen so that the point A has coordinates $(0,0)$ and so that the axis $x_1 = 0$ is not tangent to F or G at A. The curve F of equation $F(x_1, x_2) = 0$ is transformed thus:

$$F(z_1, z_1 z_2) = \Phi(z_1, z_2) = z_1^r \phi(z_1, z_2). \tag{17}$$

The curve ϕ in the z-plane is called the *reduced transform* of the curve F. In the same way one gets the reduced transform ψ of G. The curves ϕ and ψ can have one or more points of intersection α on the line $z_1 = 0$. [For the transformation (15), one can similarly define reduced transforms, etc.] Under the transformation $z \rightarrow x$ these points α all go over into A and one obtains the formula

$$(F, G)_A = rs + \sum_{\alpha} (\phi, \psi)_\alpha. \tag{18}$$

For the proof of this formula it makes no difference whether one uses the Cremona transformation (15) or the simpler transformation (16). The proof is probably most easily carried out by first of all cutting F with the various branches z of the curve G at A. The power series for such a branch z can be written as follows:

$$X_1(\tau) = a_1 \tau^v + b_1 \tau^{v+1} + \cdots$$

$$X_2(\tau) = a_2 \tau^v + b_2 \tau^{v+1} + \cdots$$

with $a_1 \neq 0$. The exponent v is called the *multiplicity* of the branch z at A. The transformation $x \rightarrow z$ carries z over into a branch z' in the z-plane having its center α on the axis $z_1 = 0$. Then one should show

$$(F, z)_A = rv + (\phi, z')_\alpha. \tag{19}$$

Summing over all branches z of the curve G at A, one obtains (18) directly from (19).

The formula (19) is a special case of a more general formula for the intersection multiplicity of a hypersurface with a branch in the projective space P^m that I proved in my work [7] "On infinitely near points." In our special case the proof runs as follows:

Let the power series expansion for the transformed branch z' be

$$\begin{aligned} Z_1(\tau) &= a_1\tau^v + b_1\tau^{v+1} + \cdots \\ Z_2(\tau) &= p + q\tau + \cdots . \end{aligned} \tag{20}$$

From this, using the transformation (16) one obtains the power series expansion of the branch z:

$$\begin{aligned} X_1 &= X_1(\tau) = Z_1(\tau) \\ X_2 &= X_2(\tau) = Z_1(\tau)Z_2(\tau). \end{aligned}$$

Substituting these in F and taking note of (17), one obtains:

$$\begin{aligned} F(X_1, X_2) &= F(Z_1, Z_1Z_2) \\ &= Z_1^r\phi(Z_1, Z_2). \end{aligned} \tag{21}$$

If the expansion of $\phi(Z_1, Z_2)$ according to powers in τ begins with τ^w, then the expansion of $F(X_1, X_2)$, as one sees from (21), begins with

$$\tau^{rv+w}.$$

From this (19) follows, and then (18).

From (18) it follows that every term on the right is smaller than the left-hand side:

$$(\phi, \psi)_\alpha < (F, G)_A. \tag{22}$$

This is all we need for the following.

4. PROOF OF THE LINEAR CONNECTEDNESS THEOREM FOR $U = P^1$ AND $U = P^2$

A. The case $U = P^1$. Let a rational mapping of the projective line P^1 on an image variety R be given by

$$\beta\eta_k = F_k(\tau_0, \tau_1), \tag{23}$$

where the F_k are forms of the same degree. If $D(\tau)$ is the greatest common divisor of the F_k, then one can place $\beta = D(\tau)$ and cancel the factor $D(\tau)$ left and right; one thus obtains

$$\eta_k = f_d(\tau_0, \tau_1). \tag{24}$$

The forms f_k now have no common divisor, hence no common zero. Therefore, every point t of the projective line is regular for the mapping, and the image point y is uniquely determined from

$$y_k = f_k(t_0, t_1). \tag{25}$$

The linear connectedness theorem is trivial in this case.

The equations of the correspondence K are

$$y_i f_k(t) - y_k f_i(t) = 0. \tag{26}$$

The image variety R has dimension 1 or 0. If the dimension is 1, then R is a rational curve with the parametric representation (25). If it is zero, then R is a single point.

B. The case $U = P^2$. Let a rational mapping of P^2 into P^m be given by

$$\beta\eta_k = F_k(\xi_0, \xi_1, \xi_2). \tag{27}$$

We can again assume that the forms F_k have no common divisor. One is to show that the image variety V_A of a point A of P^2 is linearly connected.

Let the point A have coordinates $(1, 0, 0)$. One can again place $x_0 = 1$ and introduce non-homogeneous coordinates x_1, x_2. Under the substitution $X_0 = 1$ the forms $F_k(X)$ go over into non-homogeneous polynomials $F_k(X_1, X_2)$. We write them as sums of homogeneous polynomials of ascending degree $r, r+1, \ldots$:

$$F_k = L_k + M_k + \cdots . \tag{28}$$

To get all the image points, one has to substitute into the polynomials F_k the power series expansions of all the branches z centered at A. Let the power series expansion of one such branch z be given by

$$\begin{aligned} X_1(\tau) &= a_1\tau^v + b_1\tau^{v+1} + \cdots \\ X_2(\tau) &= a_2\tau^v + b_2\tau^{v+1} + \cdots , \end{aligned} \tag{29}$$

where a_1 and a_2 are not both zero. The ratio $a_2 : a_1$ defines the *tangential direction* of the branch z.

If one now substitutes (29) in (28), then one gets

$$F_k(X(\tau)) = \tau^{rv}L_k(a_1, a_2) + \cdots . \tag{30}$$

There are only a finite number of tangential directions for which all the $L_k(a_1, a_2)$ become zero. We call them the *principal tangential directions*. For all other directions the image point is given simply by

$$y_k = L_k(a_1, a_2). \tag{31}$$

The equations (31) define a rational curve R or a single point in the image space. This curve or point belongs in any case to the image variety V_A. All other parts of V_A come from the principal tangential directions.

One can form two linear combinations of the F_k with $2(n+1)$ independent indeterminates λ_k and μ_k:

$$F_\lambda = \sum \lambda_k F_k, \tag{32}$$

$$F_\mu = \sum \mu_k F_k. \tag{33}$$

Since the F_k have no common factor, also F_λ and F_μ have none. The cycles F_λ and F_μ therefore have a definite intersection multiplicity m at the point A. If $m = 0$, then the F_k are not all zero at A. The point A is then regular, V_A is a single point and the linear connectedness of V_A is trivial. We therefore assume $m>0$ and make an induction on m. For a definite value of m the assertion that V_A is linearly connected is to be proved under the induction assumption that it is correct for all smaller values of m.

Let the coordinate system be chosen so that the axis $x_1=0$ does not coincide with a principal tangential direction. The equations

$$x_1 = s_1, \qquad x_2 = s_1 s_2 \tag{34}$$

define, as in §3, a rational mapping $s \to x$. The composition of the mapping $s \to x$ with the given rational mapping $x \to y$ yields a rational mapping $s \to y$. To a general point (S_1, S_2) of the s-plane with indeterminate coordinates S_1, S_2 there corresponds under this mapping the point η with coordinates

$$\beta\eta_k = F_k(S_1, S_1S_2) = \Phi_k(S_1, S_2). \tag{35}$$

As in §3 one can place

$$\Phi_k(S_1, S_2) = S_1^r \phi_k(S_1, S_2), \tag{36}$$

where the ϕ_k are not all divisible by S_1. From (35) and (36) there follows

$$\beta\eta_k = F_k(S_1, S_1S_2) = S_1^r\phi_k(S_1, S_2)$$

or, if one places $\beta = \gamma S_1^r$,

$$\gamma\eta_k = \phi_k(S_1, S_2). \tag{37}$$

The mapping $s \to y$ is thus effected by the polynomials ϕ_k.

Multiplying (35) and (36) with λ_k and summing over k, one obtains

$$F_\lambda(S_1, S_1S_2) = \sum \lambda_k \Phi_k(S_1, S_2) = S_1^r \sum \lambda_k \phi_k(S_1, S_2)$$

or, if the sum on the right is called ϕ_λ,

$$F_\lambda(S_1, S_1S_2) = S_1^r \phi_\lambda(S_1, S_2)$$

and likewise

$$F_\mu(S_1, S_1S_2) = S_1^r \phi_\mu(S_1, S_2).$$

Thus the cycles ϕ_λ and ϕ_μ are the reduced transforms of the cycles F_λ and F_μ in the sense of §3. Hence the intersection multiplicity m' of ϕ_λ, ϕ_μ at any point α of the axis $s_1 = 0$ is smaller than the intersection multiplicity of F_λ and F_μ at the point A. By the induction assumption the image W_α of the point α under the mapping $s \to y$ is therefore linearly connected.

Now it is easy to prove the linear connectedness of V_A. V_A is the set of images y that come from the various branches z in the x-plane having center A. The branches with non-principal direction yield points of the rational curve R. The branches z that have a definite principal direction go over, under the inverse transformation $x \to s$, into branches z' having some point α on the axis $s_1 = 0$ as center. Instead of applying the transformation $x \to y$ to the branch z, one can just as well apply the transformation $s \to y$ to the branch z': the result is the same. Hence the points y that one gets from these branches z form precisely W_α. Hence V_A is the union of the rational curve R and the finite number of linearly connected varieties W_α belonging to the principal directions.

We show now that R has a point in common with each W_α. From this the linear connectedness of V_A follows at once.

Let the axis $s_1 = 0$, complete with a point at infinity, be called g. In the mapping $s \to x$, to the individual points of g there correspond tangential directions $a_1 : a_2$ at A. To these there correspond in turn points of R. Hence the mapping $s \to y$ maps the line g onto the rational curve R. [In other words, the mapping $s \to y$ induces a rational mapping on g, and this is into R, as one sees by considering a generic point over k of g, which corresponds

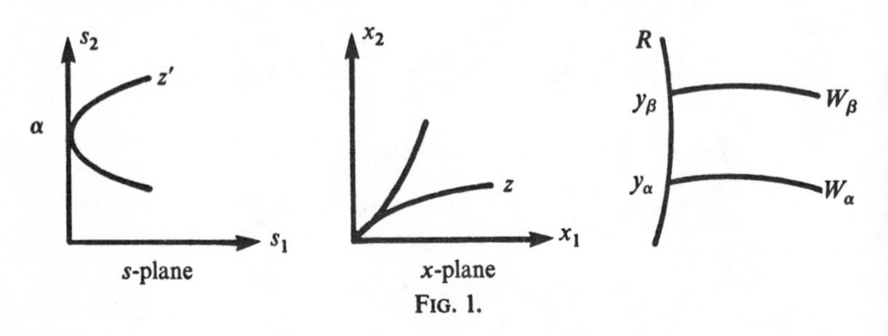

FIG. 1.

to a non-principal direction and hence to a point of R. The generic point is mapped into a generic point of R, and hence g is mapped onto R.]

If one now approaches the point α along the line g, one gets, in the mapping $s \rightarrow y$, a limit point y_α, which lies as well in R as also in W_α. Hence R has a point y_α in common with each W_α. Since R and all the W_α are linearly connected, their union V_A is linearly connected.

5. PROOF OF THE LINEAR CONNECTEDNESS THEOREM FOR $U = P^m$

Let a rational mapping of P^m into P^n be given by

$$\beta\eta_k = F_k(\xi_0, \ldots, \xi_n). \tag{38}$$

Let the point A have coordinates $(1, 0, \ldots, 0)$. We wish to show that the image variety V_A is linearly connected, that is, that any two points y' and y'' of V_A can be connected by a chain of rational curves lying on V_A.

According to §2 there are branches z' and z'' in P^m centered at A to which there correspond branches in P^n having centers y' and y''. In the power series expansions for the branch z' one can divide all the $X_i(\tau)$ by $X_0(\tau)$, so that we will have $X_0(\tau) = 1$. If one writes σ instead of τ and x_i' instead of X_i, one obtains for the branch z' the series

$$x_i'(\sigma) = a_i\sigma + b_i\sigma^2 + \cdots \qquad (i = 1, \ldots, m).$$

According to §2 one can break off the power series without changing the image point y'. One can therefore assume that

$$x_i'(\sigma) = a_i\sigma + b_i\sigma^2 + \cdots + e_i\sigma^q. \tag{39}$$

Likewise for z'':

$$x_i''(\tau) = g_i\tau + h_i\tau^2 + \cdots + k_i\tau^r. \tag{40}$$

Here σ and τ are indeterminates.

If one now forms the sum

$$x_i(\sigma, \tau) = x_i'(\sigma) + x_i''(\tau), \tag{41}$$

then this retains a meaning if one substitutes arbitrary field elements s and t for σ and τ, and one obtains a rational mapping

$$(s,t)\to x$$

of the affine (s,t)-plane into the space P^m. All points of the affine plane are regular for this mapping. In the mapping the s-axis $(t=0)$ yields the branch z', the t-axis the branch z''.

If one now composes the mapping $(s,t)\to x$ with the mapping $x\to y$, one obtains a rational mapping of the (s,t)-plane into P^n. One obtains the complete image W_Q of the point $Q:(0,0)$ in this mapping by first mapping the branches having Q as center into P^m by $(s,t)\to x$ and then into P^n by the mapping $x\to y$; the centers of the branches obtained in this way form W_Q. Obviously W_Q is contained in V_A. If one approaches the point Q along the s-axis, one obtains the point y' as image point in the y-space. If one approaches it along the t-axis, one gets the image point y''. According to §4, W_Q is linearly connected, hence one can connect y' and y'' by a chain of rational curves lying on W_Q.

Thus we have proved:

THEOREM 2: *In a rational mapping of $U=P^m$ into P^n, the image variety V_A of any point A is linearly connected.*

It is clear that the proof also holds if U is the affine space A^m.

6. THE UNIQUENESS OF THE MULTIPLICITY IN CASE $U=P^m$ OR $U=A^m$

Let a normal problem be defined by the equations

$$G_j(x,y)=0, \tag{42}$$

where the point x runs over the projective space $U=P^m$ or the affine space $U=A^m$. Let the problem have only a finite number of solutions $\eta^{(1)},\dots,\eta^{(h)}$ for a general point ξ of this space. The specialization $\xi \to x$ can be extended to a specialization

$$(\xi, \eta^{(1)}, \ldots, \eta^{(h)}) \to (x, y^{(1)}, \ldots, y^{(h)}). \tag{43}$$

We wish to prove:

THEOREM 3: *If the variety V_x of the solutions y belonging to a point x decomposes into two disjoint subvarieties M and N, then the number of specialized solutions $y^{(\nu)}$ that fall in M is uniquely determined. One can call this number the multiplicity of M.*

For the proof one forms the product

$$P(U) = \gamma \prod_{\nu} (U_0 \eta_0^{(\nu)} + \cdots + U_n \eta_n^{(\nu)}), \tag{44}$$

where the U_i are indeterminates and γ is an arbitrary constant different from zero. If the coordinates η are normalized (say by placing $\eta_s = 1$ if η_s is the first coordinate different from zero [$s = 0$ for all $\eta^{(\nu)}$ if the coordinates η are subjected to a suitable non-singular homogeneous linear transformation]), then the product on the right in (44) is uniquely determined and symmetric in the $\eta^{(1)}, \ldots, \eta^{(h)}$. If the $\eta^{(\nu)}$ are separable over $k(\xi)$, then the product is rational in the ξ. If the $\eta^{(\nu)}$ are not separable, then a qth power of the product is rational in the ξ, where $q = p^e$ is a power of the characteristic of the field k.

In the following the proofs will be carried out only for the separable case. The modifications in the inseparable case are trivial: one need only raise all the formulas to a suitable p^eth power. We assume, then, that the product $P(U)$ is rational in the ξ. By suitable choice of γ one can even arrange $P(U)$ to become a polynomial in the ξ.

Let the coefficients of the form (44) be $\zeta_0, \ldots, \zeta_N$. They are the coordinates of the cycle ζ consisting of the points $\eta^{(1)}, \ldots, \eta^{(h)}$ each counted once. The ζ_i can also be considered as coordinates of an image point ζ in a projective space P^N.

If one compares the coefficients of the power products of the U_j on the left and right in (44), then one obtains

$$\zeta_k = \gamma g_k(\eta^{(1)}, \ldots, \eta^{(h)}). \tag{45}$$

From (45) there follows

$$\zeta_j g_k(\eta) - \zeta_k g_j(\eta) = 0. \tag{46}$$

Conversely, (45) and (44) follow from (46). The equations (46) therefore express that the cycle ζ consists precisely of the points $\eta^{(1)}, \ldots, \eta^{(h)}$.

Every specialization (43) can be extended to a specialization

$$(\xi, \eta^{(1)}, \ldots, \eta^{(h)}, \zeta) \rightarrow (x, y^{(1)}, \ldots, y^{(h)}, z). \tag{47}$$

The equations (46) remain valid, so the cycle z consists precisely of the points $y^{(1)}, \ldots, y^{(h)}$. In particular it follows that

$$(\xi, \zeta) \rightarrow (x, z) \tag{48}$$

is a specialization. Conversely, every specialization (48) can be extended to a specialization (47), so every point z obtained from (48) represents a cycle, which consists of points $y^{(1)}, \ldots, y^{(h)}$. In other words:

The problem of obtaining all the specializations (43) *is precisely the same as that of obtaining all the specializations* (48).

We have seen that the coefficients of the product (44), hence the coordinates ζ_k of the cycle ζ, are rational functions of the ξ_i. Thus one has a rational mapping

$$\beta \zeta_k = F_k(\xi). \tag{49}$$

The pairs (x, z) of this mapping are precisely the specializations of the general pair (ξ, ζ). So the problem of obtaining all the systems of solutions $y^{(1)}, \ldots, y^{(h)}$ for given x is reduced to the problem of obtaining all the image points z for a given point x in the rational mapping (49). According to §5 the set W_x of image points z is a linearly connected variety.

The hypothesis of Theorem 3 says that the variety V_x of the solutions y belonging to x decomposes into two disjoint subvarieties M and N. Suppose now that the number of points $y^{(\nu)}$ that fall in M were not uniquely determined. Then there would be a specialization $y^{(1)}, \ldots, y^{(h)}$ for which only r of the points $y^{(\nu)}$ would fall in M and the other $h-r$ in N, while in another specialization at least $r+1$ points $y^{(\nu)}$ fall in M.

I define a correspondence K between $y^{(1)}, \ldots, y^{(h)}$ and z by the following equations:

(a) the equations $H_j(x, y^{(1)}, \ldots, y^{(h)}) = 0$ that express that

$$(\xi, \eta^{(1)}, \ldots, \eta^{(h)}) \rightarrow (x, y^{(1)}, \ldots, y^{(h)})$$

is a specialization. In these equations x is fixed; so they are equations in $y^{(1)}, \ldots, y^{(h)}$ alone. Amongst these equations also appear the equations (42) with $y = y^{(1)}$ or $\cdots$ or $y = y^{(h)}$;

(b) the equations

$$z_j g_k(y^{(1)}, \ldots, y^{(h)}) - z_k g_j(y^{(1)}, \ldots, y^{(h)}) = 0 \tag{50}$$

analogous to (46), which say that the cycle z consists precisely of the points $y^{(1)}, \ldots, y^{(h)}$.

The correspondence K is the union of two disjoint sub-correspondences K_1 and K_2. Here K_1 consists of the systems $(y^{(1)}, \ldots, y^{(h)}, z)$ in which at least $r+1$ points $y^{(\nu)}$ lie in M and K_2 consists of the systems in which at least $n-r$ points $y^{(\nu)}$ lie in N. That K_1 and K_2 are disjoint is clear. We still have to show that K_1 and K_2 are correspondences, i.e., that they can be defined by algebraic equations.

We choose a combination κ of $r+1$ points $y^{(\nu)}$ and adjoin to the equations (a) and (b) the equations that say that these points belong to M. Thus one obtains a correspondence $K_1^{(\kappa)}$ for every κ. The union of these correspondences is K_1. Similarly for K_2. Therefore one has in fact a decomposition of K into two disjoint sub-correspondences K_1 and K_2.

K, K_1, and K_2 are correspondences between the h-fold projective space $P^{n,n,\ldots,n}$ of systems of points $(y^{(1)}, \ldots, y^{(h)})$ and the projective space P^N of the points z. The image varieties W, W_1, and W_2 belonging, respectively, to K, K_1, and K_2 consist of all the points z of P^N to which belong systems $(y^{(1)}, \ldots, y^{(h)}, z)$ of K, K_1, and K_2, respectively. Since K is the union of K_1 and K_2, W is the union of W_1 and W_2. These are disjoint, since a cycle of only h points cannot have $r+1$ points in M and $n-r$ points in N. Hence W decomposes into two disjoint subvarieties W_1 and W_2.

On the other hand, W consists precisely of the points z that one gets from the specializations (48). Namely, every such specialization can be extended to a specialization of (47), as was previously observed. So W is precisely the image variety of the point x in the rational mapping (49), previously denoted by W_x. This is connected, however, so it cannot decompose into disjoint sub-varieties.

Thus Theorem 3 is proved.

If V_x decomposes into the maximal connected subvarieties $M_1, \ldots, M_t$, then by Theorem 3 each M_i has a definite multiplicity μ_i. The sum of these multiplicities is equal to the number of solutions in the general case:

$$h = \mu_1 + \cdots + \mu_t.$$

This statement is called "the principle of the conservation of number."

As a special case of Theorem 3 one obtains:

THEOREM 4: *The multiplicity of an isolated solution y is uniquely defined.*

If in particular a normal problem consists in cutting a d-dimensional irreducible variety with d hyperplanes, which are considered as specializations of general hyperplanes, then by Theorem 4 each isolated intersection y has a uniquely defined multiplicity. If A is a given point, one passes d hyperplanes through A in the most general way possible and cuts V with them. If the multiplicity of the intersection A is equal to one, then A is called a *simple point* of V.

7. INTERSECTION MULTIPLICITIES

If two irreducible varieties V^r and V^{n-r} on the irreducible variety W^n have an isolated point of intersection A that is simple for W^n, then one can define the intersection multiplicity of V^r and V^{n-r} at A either according to Severi and van der Waerden [8] or according to Weil [2]. The first definition was given originally only for the case that V^r and V^{n-r} have just a finite number of points of intersection, but on the basis of the general concept of intersection multiplicity here established in §6 one can extend the definition immediately to the case of an isolated point of intersection. Weil's definition was from the beginning adapted to this case.

Severi and van der Waerden proceed in two steps. First it is assumed that V^r and V^{n-r} lie in a projective space. Then V^r is brought into a general position relative to V^{n-r} by a projective transformation with indeterminate coefficients τ_{ik}. If

$$F_\nu(x_0, \ldots, x_n) = 0$$

are the equations for V^r, then

$$F_\nu\left(\sum \tau_{ik} x_k\right) = 0$$

are the equations for the transformed variety $(V^r)^T$. This cuts V^{n-r} in a finite number of points. If the matrix $T = (\tau_{ik})$ is specialized to the identity matrix, the isolated point of intersection A gets a definite multiplicity. This is defined as the intersection multiplicity of V^r and V^{n-r} at A.

Second, let V^r and V^{n-r} be varieties on W^n, where W^n is contained in a projective space P^N. If one joins the points of V^r by lines with the points of a general linear sub-space L^{N-n-1} of P^N, one obtains a projecting cylinder K^{N+r-n}. The multiplicity of A as intersection of K^{N+r-n} and V^{n-r} is now defined as the intersection multiplicity of V^r and V^{n-r} at A.

Weil also proceeds in two steps. First he defines the intersection multiplicity of an isolated point of intersection of V^r with a linear subspace L^{n-r} of affine space A^n by considering L^{n-r} as specialization of a general linear space. If now, second, an isolated point of intersection A of V^r and V^{n-r} on W^n is given, where W^n lies in the affine space A^N and the point A is a simple point of W^n, then Weil forms the product $V^r \times V^{n-r}$ in $A^N \times A^N$. The point $A \times A$ is an isolated point of intersection of $V^r \times V^{n-r}$ with the diagonal Δ of $A^N \times A^N$. This diagonal is a linear space in $A^N \times A^N$ given by the equations

$$X_k - X'_k = 0 \qquad (k = 1, \ldots, N).$$

Weil forms n linear combinations

$$F_i(X - X') = 0 \qquad (i = 1, \ldots, n) \tag{51}$$

of these equations, where the linear forms F_i are so chosen that the linear space L^{N-n} of A^N defined by the equations $F_i(X - A) = 0$ intersects the tangent space of W^n at A only in A. Weil calls such a system of n linear forms "a uniformizing set of linear forms for W at A." If the linear forms F_i are thus chosen, then the equations (51) define a linear subspace of $A^N \times A^N$ having the point $A \times A$ as isolated point of intersection with $V^r \times V^{n-r}$. The multiplicity of this point of intersection is defined as the intersection multiplicity of V^r and V^{n-r} at A.

Leung [4] has shown that the definition of Severi and van der Waerden is equivalent with that of Weil.

Chevalley [9] gave another definition of intersection multiplicity. Samuel [10] has shown that the definition of Chevalley is equivalent with that of Weil.

8. THE LINEAR CONNECTEDNESS THEOREM FOR SIMPLE POINTS OF ARBITRARY VARIETIES U

Using a parallel projection, one can reduce the linear connectedness theorem for a simple point x on U to the special case $U = P^m$. The method is the same as that applied in an earlier work [11] in the proof of the uniqueness of the specialization multiplicity.

Let the point x have the homogeneous coordinates

$$x_0 = 1, x_1 = 0, \ldots, x_N = 0.$$

For a general point ξ of U one has $\xi_0 \neq 0$; one can therefore place $\xi_0 = 1$ and introduce non-homogeneous coordinates $\xi_1, \ldots, \xi_N$. Let a rational mapping $U \to V$ be given by

$$\beta\eta_k = F_k(\xi_1, \ldots, \xi_N). \tag{52}$$

We are to show that the image variety V_x of the simple point x is linearly connected.

Let the groundfield k be perfect. If one adjoins N^2 indeterminates u_{ik}, one obtains a new groundfield

$$K = k(u_{ik}).$$

Applying the linear transformation of coordinates

$$\xi_i^* = \sum u_{ik}\xi_k,$$

one finds that $\xi_1^*, \ldots, \xi_m^*$ are algebraically independent over K and that all the ξ_i^* are separable functions of $\xi_1^*, \ldots, \xi_m^*$. For the proof see, say, [12, §6]. Now one can project the entire affine space A^N, and in particular the variety U, into the subspace A^m spanned by the first m coordinates by keeping the first m coordinates

$$\tau_1 = \xi_1^*, \ldots, \tau_m = \xi_m^*$$

and replacing the others by zero or, better, leaving them aside entirely. The projection is therefore defined by the formulas

$$\tau_i = \sum_1^N u_{ik}\xi_k \qquad (i = 1, \ldots, m). \tag{53}$$

The τ_i are algebraically independent over K, hence can be thought of as indeterminates. The ξ_k are linear functions of the ξ_i^*, hence are separable algebraic functions of $\tau_1, \ldots, \tau_m$.

Now for given τ we seek those points X of U that project into τ. They must satisfy the equations of U and the linear equations

$$\sum u_{ik}X_k = \tau_i \tag{54}$$

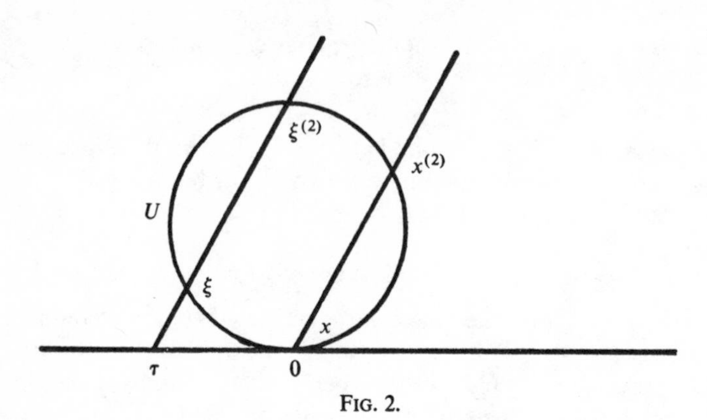

FIG. 2.

or, written homogeneously,

$$\sum u_{ik} X_k - \tau_i X_0 = 0. \tag{55}$$

There are no points at infinity satisfying (55) and lying on U, since the $(m-1)$-dimensional intersection of U with the hyperplane at infinity $X_0 = 0$ has no point in common with m general hyperplanes $\sum u_{ik} X_k = 0$. Hence we can again place $X_0 = 1$ in (55) and revert to the non-homogeneous equations (54).

Every solution X of (54) that lies on U has the maximal degree of transcendency m, since by (54) the m independent field elements $\tau_1, \ldots, \tau_m$ lie in the field $K(X)$. Hence the X are general points of U over K. Therefore for each point X there is an isomorphism over K that carries ξ into X. Because of (53) and (54) this isomorphism leaves the elements τ_i fixed. Hence there are only a finite number of points X, namely, the points

$$\xi = \xi^{(1)}, \xi^{(2)}, \ldots, \xi^{(g)}$$

conjugate to ξ.

The number g is the degree of U: the number of points of interesection of U with m general hyperplanes (55). All symmetric functions of $\xi^{(1)}, \ldots, \xi^{(g)}$ are rational in the τ_i.

In the mapping (52) all the points $\xi = \xi^{(1)}, \ldots, \xi^{(g)}$ have definite images $\eta = \eta^{(1)}, \ldots, \eta^{(g)}$. All symmetric functions of

$\eta^{(1)}, \ldots, \eta^{(g)}$, in particular the coordinates ζ_k of the cycle formed of these points, are rational functions of $\tau_1, \ldots, \tau_m$:

$$\gamma\zeta_k = G_k(\tau_1, \ldots, \tau_m). \tag{56}$$

To obtain all the images y of the simple point x in the mapping (52), one must extend the specialization $\xi \to x$ in all possible ways to specializations

$$(\xi, \eta) \to (x, y). \tag{57}$$

Under the specialization $\xi \to x$ the τ_i defined by (53) go over, of course, into zero. Hence one can extend only in such a way that

$$(\tau, \xi, \eta) \to (0, x, y). \tag{58}$$

Here 0 is the point in the τ-space with coordinates zero.

Then one can extend (58) to

$$\begin{aligned}(\tau, \xi, \xi^{(2)}, \ldots, \xi^{(g)}; \eta, \eta^{(2)}, \ldots, \eta^{(g)}, \zeta)& \\ \to (0, x, x^{(2)}, \ldots, x^{(g)}; y, y^{(2)}, \ldots, y^{(g)}, z).&\end{aligned} \tag{59}$$

The equations expressing that the cycle ζ consists of the points $\eta = \eta^{(1)}, \ldots, \eta^{(g)}$ continue to hold under specialization, so the cycle z consists of the points $y = y^{(1)}, \ldots, y^{(g)}$. The point $\eta^{(\nu)}$ is an image of $\xi^{(\nu)}$ in the mapping (52), so $y^{(\nu)}$ is an image of $x^{(\nu)}$ in this same mapping. The points $x^{(1)}, \ldots, x^{(g)}$ arise from $\xi^{(1)}, \ldots, \xi^{(g)}$ under specialization, hence are the points of intersection of U with the linear space

$$\sum u_{ik} X_k = 0. \tag{60}$$

Since x is a simple point of U, x occurs only once among $x^{(1)}, \ldots, x^{(g)}$, i.e., $x^{(2)}, \ldots, x^{(g)}$ are different from $x^{(1)} = x$. Now we assert:

The points $x^{(2)}, \ldots, x^{(g)}$ are regular for the mapping (52).

Proof: The non-regular points on U form a subvariety of dimension at most $m - 1$. If one cuts this successively by the m

hyperplanes (60), which pass through x, the dimension goes down 1 at each of the first $m-1$ steps. After $m-1$ steps one obtains a finite number of points of intersection, $x, x', x'', \ldots$. Taking one more step with the mth hyperplane, one is left only with the point x. Hence the points $x^{(2)}, \ldots, x^{(g)}$ differing from x are regular and their images $y^{(2)}, \ldots, y^{(g)}$ are uniquely determined. Only the one image point y can vary, and indeed it runs precisely through the image variety V_x of the point x. From this there follows at once:

To every point y of V_x there belongs a definite cycle

$$z = y + y^{(2)} + \cdots + y^{(g)} \tag{61}$$

such that the associated point z lies in the image variety W_O of the point O in the rational mapping (56).

The converse also holds:

To every point z of W_O there belongs a cycle z having precisely the form (61), *with y lying in V_x.*

Proof: Every specialization $(\tau, \zeta) \to (0, z)$ can be extended to a specialization (59). Here it is possible that not $\xi = \xi^{(1)}$ but some other $\xi^{(\nu)}$ specializes to x. Since, however, the $\xi^{(\nu)}$ are all conjugate to ξ, one can apply an automorphism of the field

$$K(\xi^{(1)}, \ldots, \xi^{(g)})$$

to carry $\xi^{(\nu)}$ into ξ and thus arrange that under the specialization ξ goes over into x and the other $\xi^{(\nu)}$ into $x^{(2)}, \ldots, x^{(g)}$ in some order. Then $\eta^{(2)}, \ldots, \eta^{(g)}$ also go over into the uniquely determined points $y^{(2)}, \ldots, y^{(g)}$, and the cycle z has the form (61).

Hence (61) defines a one to one mapping between V_x and W_O: To each y in V_x there corresponds a z in W_O and conversely.

This mapping is a projectivity.

Proof: According to §6 the coordinates z_i of the cycle z are the coefficients of the product

$$P(U) = r\left(\sum U_k y_k\right) \cdot \prod_{2}^{g} \left(\sum U_k y_k^{(\nu)}\right),$$

hence the z_i are linear homogeneous functions of the y_k. Therefore the mapping $y \to z$ is a projectivity.

Now W_O is linearly connected, hence V_x also is linearly connected, which was to be proved.

REFERENCES

1. B. L. van der Waerden, "Der Multiplizitätsbegriff der algebraischen Geometrie," *Math. Ann.*, **97** (1927), 756.
2. A. Weil, *Foundations of Algebraic Geometry*, Amer. Math. Soc. Coll. Publications, vol. 29, 1st ed., 1946; 2nd ed., 1962.
3. D. Northcott, "Specializations over a local domain," *Proc. London Math. Soc.*, **1** (3) (1951).
4. K.-T. Leung, "Die Multiplizitäten in der algebraischen Geometrie," *Math. Ann.*, **135** (1958), 170.
5. O. Zariski, "Theory and applications of holomorphic functions on algebraic varieties," *Mem. Amer. Math. Soc.*, **5** (1951).
6. W.-L. Chow: "On the connectedness theorem in algebraic geometry," *Amer. J. Math.*, **81** (1959), 1033.
7. B. L. van der Waerden, "Infinitely near points," *Proc. Akad. Amsterdam*, **53** (1950), 401.
8. ———, "Zur algebraischen Geometrie," 14, *Math. Ann.*, **115** (1938), 621.
9. C. Chevalley, "Intersection of algebraic and algebroid varieties," *Trans. Amer. Math. Soc.*, **57** (1945), 1.
10. P. Samuel, "La notion de multiplicité," *J. Math. Pures Appl.*, **30** (1951), 159.
11. B. L. van der Waerden, "Zur algebraischen Geometrie," 6, §3, *Math. Ann.*, **110** (1934), 144.
12. ———, "Verallgemeinerung des Bézoutschen Theorems," *Math. Ann.*, **99** (1928), 497.

SPACE CURVES AS IDEAL-THEORETIC COMPLETE INTERSECTIONS

Jack Ohm

The question of how many polynomials are needed to generate a given prime ideal p in a polynomial ring $R = k[X_1, \ldots, X_n]$, with, say, k an algebraically closed field, has been around a long time, although originally in a more geometric guise. Two cases are immediately noteworthy: the ht 1 primes of R are principal, as is so with any UFD; and the maximal primes of R have the form $(X_1 - \alpha_1, \ldots, X_n - \alpha_n)$, $\alpha_i \in k$, by Hilbert's Nullstellensatz. The first example not covered by these cases is that of a ht 2 prime in $k[X_1, X_2, X_3]$, or, in other words, the defining ideal of an irreducible (affine) space curve; and it was Macaulay who showed that there exist primes of this type which require arbitrarily large numbers of generators. (Only recently have all the details of Macaulay's argument, which undoubtedly has confounded generations of graduate students, been written out, by Abhyankar and Geyer.)

Since the Macaulay examples are space curves with a singularity at the origin, subsequent research has focused on non-singular curves and, in particular, on Serre's question of whether a non-singular irreducible space curve of genus $\leqslant 1$ is a complete intersection, i.e., whether the defining prime ideal of such a curve can be generated by 2 elements. Serre pointed out that it would suffice to prove that finitely generated rk 2 projective modules over $k[X_1, X_2, X_3]$ are free, and it was in this setting that the question was recently settled in the affirmative by Murthy–Towber. In fact, Quillen and, independently, Suslin have now provided an elementary solution to the general Serre problem of proving that finitely generated projectives over $k[X_1, \ldots, X_n]$ are free (for any PID k). (See [58] and [21].)

We shall be concerned below not with the Serre problem itself, but rather with the relationship between it and minimal numbers of generators, and with applications to space curves. The work is divided into five sections:

I. *The geometric setting.* The requirement that the defining ideal of a non-singular irreducible space curve C be generated by 2 elements is somewhat stronger than the requirement that the point-set of C be exactly the intersection set of two surfaces, and this section is devoted to giving a geometric interpretation of the distinction.

II. *A local-global principle.* The aim here is to explain how the assumption that finitely generated projectives are free may be used to gain information on minimal numbers of generators. As an application, it is shown that rk 3 projectives are free over $k[X_1, X_2, X_3]$ implies that the ideal of a non-singular space curve is generated by $\leqslant 3$ elements.

III. *When is $d(I) \leqslant 1$?* A crucial hypothesis in applying the techniques of section II to an ideal I is that the projective dim of I should be $\leqslant 1$. Section III surveys the ideals for which this is valid.

IV. *An isomorphism relating* $\mathrm{Ext}_R^1(I, R)$ *and the module of differentials.* A fundamental isomorphism connecting the module of differentials of a non-singular space curve C with a module that arises in section II is established, and then, by interpreting the genus of C in terms of the module of differentials, it is a simple matter to conclude (in section V) that a curve C of genus 0 (or 1) is a complete intersection.

V. *The module of differentials.* The definition and basic properties of the module of differentials, together with its relationship to the notion of simple point, are discussed. These facts are needed partly for the applications already outlined in section IV and partly for further applications to complete intersections given at the end of section V.

Kaplansky's book *Commutative Rings* [32] has just about the right blend of ring theory and homological algebra for our purposes, so we shall use it as a basic reference and shall assume a certain amount of familiarity with it. There are two other references that should be mentioned at the start: Geyer's excellent lecture notes [23], to which the present work may be regarded as complementary, and the very readable Russell–Sathaye notes of Abhyankar's Montreal lectures [1], which do without the use of homological methods. (The Abhyankar–Sathaye notes [3] form a sequel to [1] but are more technical.)

The following are some variations on Kaplansky's terminology that will be in effect:

1. For an ideal $I \neq R$ of a ring R, $\operatorname{ht} I$ is used instead of $\operatorname{rk} I$. Also, $\operatorname{coht} I = \operatorname{Krull\,dim} R/I$.

2. If M is a f.g. (finitely generated) R-module, $\mu(M) =$ least number of elements needed to generate M. (It will be convenient to regard the zero module as being generated by 0 elements.) If R is quasi-local, then $\mu(M)$ also equals the number of elements in any generating set which cannot be shortened.

3. Krull's principal ideal theorem (PIT) [32, p. 104] asserts that for a noetherian R, $\operatorname{ht} I \leqslant \mu(I)$; if $=$ holds, I will be called a *complete intersection* (abbreviated c.i.) *ideal*. More precisely, the PIT asserts that $\operatorname{ht} p \leqslant \mu(I)$ for any minimal prime p of I.

4. *Regular element* = non-zero-divisor. *Regular ideal* = ideal containing a regular element (and *not* an ideal generated by a regular sequence!). $\mathfrak{Z}(R) = \{$zero-divisors of $R\}$.

5. $\operatorname{Ass} I = \{$associated prime ideals of $I\}$. A prime ideal p of R is called an associated prime of the ideal I if p is a minimal prime

of $I : r$ for some $r \in R$. I is said to be *unmixed* if all elements of $\operatorname{Ass} I$ have the same ht. Note also that $\mathfrak{Z}(R/I) = \cup\{p \in \operatorname{Ass} I\}$.

6. $\operatorname{Spec} R = \{\text{prime ideals of } R\}$. $V(I) = \{p \in \operatorname{Spec} R \mid p \supset I\}$. The collection of all $V(I)$, I an ideal of R, constitutes the closed sets for the Zariski topology of $\operatorname{Spec} R$. $\sqrt{I} = \cap\{p \in V(I)\}$ and ${}^{j}\sqrt{I} = \cap\{\text{maximal ideals} \supset I\}$. R is called *reduced* if $\sqrt{(0)} = (0)$, and $R_{\text{red}} = R/\sqrt{(0)}$.

I. THE GEOMETRIC SETTING

There are two senses in which an algebraic space curve C may be thought of as being cut out completely by surfaces: the set-theoretic, where C should be exactly the point-set common to the surfaces; and the ideal-theoretic, where the defining ideal for C should be generated by the polynomials that define the surfaces. Since the latter sense lacks the intuitive geometric content of the former, it is perhaps well to begin by analyzing just what it means geometrically for a space curve to be an ideal-theoretic complete intersection of two surfaces. This will be taken up in §§6–8, but first it is necessary to develop the basic properties of simple points (§§1–5).

Fix an algebraically closed field k and a polynomial ring $k[X] = k[X_1, \ldots, X_n]$. If $I \neq k[X]$ is an ideal of $k[X]$, we shall write $k[X]/I = k[x]$, where x_i denotes the canonical image of X_i in $k[X]/I$.

1. Affine varieties. Since k is algebraically closed, there is a 1-1 correspondence between the n-tuples $(\alpha) \in k^n$ and the maximal ideals m of $k[X]$ (cf. [32, p. 19]); and (α) is a zero of the ideal I if and only if its corresponding maximal ideal $m(\alpha)$ contains I. We shall loosely use the word "point" to refer to either (α) or $m(\alpha)$, leaving it to the context to make our intent clear. The set of all such points, in one sense or the other, will be called *affine n-space* (over k) and will be denoted $\mathcal{A}^n$.

A *variety* in affine n-space will be taken to be the set $\mathcal{V}(I)$ of all maximal ideals of $k[X]$ containing a given radical ideal $I \neq k[X]$. Since every radical ideal in a polynomial ring is the intersection of the maximal ideals containing it (cf. [32, p. 18]), restricting attention to radical ideals ensures that $\mathcal{V}(I) = \mathcal{V}(J)$ implies $I = J$. On the other hand, classically a variety $\mathcal{V}$ was just the set of zeros of a collection of polynomials $f_i \in k[X]$; and since the largest collection which defines $\mathcal{V}$ in this way is the radical of the ideal generated by the f_i (cf. [32, p. 19]), nothing is lost in restricting attention to radical ideals. The (reduced) ring $k[X]/I$ is called the *affine ring* of $\mathcal{V}(I)$; and two varieties, possibly in different affine spaces over k, are said to be *isomorphic* if their affine rings are k-isomorphic. The *dimension* of $\mathcal{V}(I)$ is defined to be the Krull dimension of its affine ring, which equals $n - \operatorname{ht} I$.

2. Irreducible components. A variety $\mathcal{V}(I)$ is called *irreducible* if it is not the union of two proper subvarieties, or equivalently, if I is prime. Corresponding to the algebraic property that every radical ideal in a noetherian ring is the intersection of its finitely many minimal primes is the geometric property that every variety is the union of its finitely many maximal irreducible subvarieties, called its *irreducible components*. A variety is called *unmixed* if all its irreducible components have the same dimension. If $m \in \mathcal{V}(I)$, only the minimal primes of I contained in m are preserved in passing to the local ring $k[X]_m$; so replacing I by $I_m = Ik[X]_m$ amounts to ignoring those components of $\mathcal{V}(I)$ which do not contain m. In particular, m lies on a unique component of $\mathcal{V}(I)$ if and only if I_m is prime, or, by the commutativity of localization and homomorphic image, if and only if $k[x]_m$ is a domain. (Here $k[x]_m$ denotes $k[X]/I$ localized at the canonical image of the maximal ideal m.)

3. Simple points. A point m of the variety $\mathcal{V} = \mathcal{V}(I)$ will be called *simple* (or non-singular) on $\mathcal{V}$ if the local ring $k[x]_m$ is regular. (Recall that a local ring is called regular if its maximal ideal is a c.i. ideal, and that in general a noetherian ring is called regular if every localization at a maximal ideal is regular.) Since a regular local ring is a domain (cf. 4.1 below), by the above remarks a point of $\mathcal{V}$ is simple if and only if it lies on only one irreducible component of $\mathcal{V}$ and is simple on that component. The variety $\mathcal{V}$

will be called *non-singular* if every point of $\mathcal{V}$ is simple, or, in other words, if the affine ring $k[x]$ is regular.

There is another "classical and time-honored definition of simple point" (Zariski [74]); namely, a point $m(\alpha)$ of $\mathcal{V}$ should be simple if the Jacobian matrix of $\mathcal{V}$ at (α) has rank equal to $\operatorname{ht} I_m$ ($= \operatorname{ht} I$ if $\mathcal{V}$ is unmixed). This Jacobian matrix is defined as follows: Let $I = (f_1, \ldots, f_t)$ and let $\partial f_i/\partial \alpha_j$ denote $\partial f_i/\partial X_j$ evaluated at (α). Then $\mathcal{J}(f_i; \alpha)$ is the $t \times n$ matrix $(\partial f_i/\partial \alpha_j)$. The rank of $\mathcal{J}(f_i; \alpha)$ is independent of the choice of generators f_i for I and is $\leqslant \operatorname{ht} I_m$, as we shall see from the remarks below and Theorem 4.5; so in speaking of rank $\mathcal{J}(f_i; \alpha)$ it is not necessary to refer to the particular generators f_i for I, and we shall therefore merely write "rank $\mathcal{J}(I; \alpha)$" or "rank $\mathcal{J}(\alpha)$." (Cf. also §33.)

When (α) is the origin (which we may assume by rewriting the polynomials of $k[X]$ as polynomials in $k[X - \alpha]$), the Jacobian matrix is particularly easy to describe: $\mathcal{J}(f_i; 0) = (a_{ij})$, where $f_i = a_{i1}X_1 + \cdots + a_{in}X_n + (\text{terms of deg} \geqslant 2)$. Moreover, since the rows of $\mathcal{J}(f_i; 0)$ are the vectors $(a_{i1}, \ldots, a_{in})$, rank $\mathcal{J}(f_i; 0)$ may be characterized as the maximal number ν of elements from among $f_1, \ldots, f_t$ that are linearly independent mod m^2 over $k[X]/m = k$, where $m = (X_1, \ldots, X_n)$, or, equivalently, by Nakayama's lemma, as the maximal number of elements from I_m that form part of some minimal generating set for the ideal $mk[X]_m$. (Caution: It is important to pass to the local ring $k[X]_m$, where Nakayama's lemma is available, in discussing minimal generating sets.)

4. The number $\nu(I)$. We shall now turn to a proof of the equivalence of the two definitions of simple point given above. Aside from its interest in the present context, the algebra involved will also be useful later.

4.1. THEOREM: *Let R, m be a local ring. Then $p = (a_1, \ldots, a_t)$ is a prime ideal of* $\operatorname{ht} t$ *(if and) only if $(0) < (a_1) < (a_1, a_2) < \cdots < p$ is a chain of prime ideals (necessarily saturated).*

The proof involves two lemmas.

4.2. LEMMA: *Let R be a ring and (a) be a principal prime ideal of R. Then $(0) < (a)$ is a saturated chain of prime ideals if and only if* (i) $\mathrm{ht}(a) = 1$ *and* (ii) $\bigcap_{s=1}^{\infty}(a)^s = (0)$.

Proof: $\Rightarrow$: Assertion (i) is immediate, and (ii) follows from the fact that $\cap (b)^s = (0)$ if b is an element in a ht 1 prime of a domain [53, p. 323].

$\Leftarrow$: Since (a) is a prime of ht 1, there exists a prime Q of R such that $Q < (a)$. For any such prime Q, $Q = Q(a) = Q(a)^2 = \cdots \subset \bigcap_{s=1}^{\infty}(a)^s = (0)$, this last equality following from (ii). Therefore (0) is prime; and since $Q = (0)$ for *any* prime Q of R such that $Q < (a)$, the chain $(0) < (a)$ is saturated.

4.3. LEMMA: *Let p be a prime ideal of a noetherian ring R, let $a \in p$, and let $'$ denote canonical image in $R/(a)$. Then* $\mathrm{ht}\, p' \geqslant \mathrm{ht}\, p - 1$.

Proof: If $\mathrm{ht}\, p' = s$, then it is easily seen that p' is a minimal prime of an ideal generated by s elements: choose $b_1 \in p' \setminus \cup$ {minimal primes of (0)}, $b_2 \in p' \setminus \cup$ {minimal primes of (b_1)}, etc.; and then p' is a minimal prime of $(b_1, \ldots, b_s)$. By taking inverse images, we conclude that p is a minimal prime of an ideal generated by $s + 1$ elements. Hence by Krull's PIT, $\mathrm{ht}\, p \leqslant s + 1$.

Proof of theorem 4.1: Proceed by induction on t. If $t = 1$, then the theorem follows from 4.2 since 4.2 (ii) holds in any local ring by Krull's intersection theorem [32, p. 51]. If $t > 1$, let $R' = R/(a_1)$, and apply 4.3 to conclude $\mathrm{ht}\, p' \geqslant t - 1$. Since $p' = (a_2', \ldots, a_t')$, then $\mathrm{ht}\, p' = t - 1$ by the PIT. Now apply the induction hypothesis to p' and take inverse images to conclude $(a_1) < (a_1, a_2) < \cdots < p$ is a chain of primes.

It remains to show (0) is prime. By the PIT, $\mathrm{ht}(a_1) \leqslant 1$; and if $\mathrm{ht}(a_1) = 1$, then (0) is prime by the $t = 1$ case of the theorem. Therefore we may assume $\mathrm{ht}(a_1) = 0$. Choose r not in (a_1) but in all other ht 0 primes of R, and let $a_1^* = a_1 + ra_2$. Since $a_2 \notin (a_1)$, $a_1^* \notin$ any ht 0 prime of R; and therefore $\mathrm{ht}(a_1^*) = 1$. Moreover,

$(a_1, a_2, \ldots, a_t) = (a_1^*, a_2, \ldots, a_t)$; so by the above argument (a_1^*) is prime. The $t = 1$ case now yields (0) is prime. Q.E.D.

Note that 4.1 includes the fact that a regular local ring is a domain; if one is willing to assume this, the above proof may be shortened.

Next let us define for an ideal $I \neq R$ of a local ring R, m a non-negative integer $\nu(I)$, which in the above discussion was seen to equal the rank of the Jacobian matrix.

4.4. DEFINITION: $\nu(I) = \sup\{s \mid I$ contains s elements which form part of a minimal generating set for $m\}$.

4.5. THEOREM: *Let R, m be a regular local ring (e.g., a localization of $k[X]$) and $I \neq R$ be an ideal of R. Then $\nu(I) \leqslant \operatorname{ht} I$, and equality holds if and only if R/I is regular. Moreover, when equality holds, then I is prime, $\nu(I) = \operatorname{ht} I = \mu(I)$, and any minimal generating set for I can be extended to a minimal generating set for m.*

Proof: Set $\nu = \nu(I)$. Then there exist $a_1, \ldots, a_\nu \in I$ and $b_{\nu+1}, \ldots, b_t \in m$ such that $a_1, \ldots, a_\nu, b_{\nu+1}, \ldots, b_t$ is a minimal generating set for m. Since R is regular, $\operatorname{ht} m = t$; and hence by 4.1, $(0) < (a_1) < (a_1, a_2) < \cdots < (a_1, \ldots, a_\nu)$ is a saturated chain of primes. But $(a_1, \ldots, a_\nu) \subset I$, so therefore $\nu \leqslant \operatorname{ht} I$. Moreover, if $\nu = \operatorname{ht} I$, we can conclude $I = (a_1, \ldots, a_\nu)$, in which case I is prime and $\nu = \operatorname{ht} I = \mu(I)$ (since by Krull's PIT we always have $\operatorname{ht} I \leqslant \mu(I)$). Also, if $\nu = \operatorname{ht} I$, it follows from 4.1 and the PIT that $t - \nu \leqslant \operatorname{ht}(m/I) \leqslant \mu(m/I) \leqslant t - \nu$; and thus R/I is regular.

To complete the proof, it remains to show that if R/I is regular, then any minimal generating set $a_1, \ldots, a_s$ for I extends to a minimal generating set for m. Let $b'_1, \ldots, b'_t$ be a minimal generating set for m/I, and select pre-images b_i for the b'_i. Then $a_1, \ldots, a_s, b_1, \ldots, b_t$ is a generating set for m and hence contains a minimal generating set $\mathbb{S}$ for m. Since $b'_1, \ldots, b'_t$ is a minimal generating set for m/I, $\mathbb{S}$ necessarily contains $b_1, \ldots, b_t$; and therefore by renumbering the a_i, we may assume $\mathbb{S} =$

$\{a_1, \ldots, a_r, b_1, \ldots, b_t\}$, $r \leqslant s$. But the chain of ideals

$$(0) < (a_1) < \cdots < (a_1, \ldots, a_r) \subset I < I + (b_1) < \cdots$$
$$< I + (b_1, \ldots, b_t) = m$$

has its first $r+1$ members prime by 4.1 applied to m and its last $t+1$ members prime by 4.1 applied to m/I. Since $\operatorname{ht} m = \mu(m) = r + t$, it follows that $(a_1, \ldots, a_r) = I$. Thus, $r = s$. Q.E.D.

4.6. COROLLARY: *If $I \neq R$ is an ideal of a regular ring R such that R/I is regular, then I is a radical ideal and is locally a* c.i. *at primes $\supset I$ (i.e., $\operatorname{ht} I_p = \mu(I_p)$ for every prime $p \supset I$).*

Note that the converse to 4.6 is false: the prime ideal $I = (Y^2 - X^3)$ in $k[X, Y]$ is locally a c.i. at primes $\supset I$, but $k[X, Y]/I$ is not regular.

REMARK: Much of the above originated in the fundamental paper [74] of Zariski, which remains the most readable account of simple points that I know of. Cohen's thesis [14] contains a section on the algebraic properties of regular local rings, and the $p = m$ case of 4.1 (which is all that is used in the proof of 4.5) can be found there. Theorem 4.1 in the generality presented here is due to Davis (cf. [15]) and was made known to me by D. D. Anderson. Lemma 4.2 is an attempt on my part to pinpoint the ingredients of 4.1 for an arbitrary ring. Note also that, in the spirit of 4.1, the assertions of 4.5 generalize to the case that m (and m/I in the if and only if part) is a c.i. prime of a local ring R; the same proof works. Question: If p is a c.i. prime ideal of a noetherian ring R, is it always possible to find some minimal generating set $a_1, \ldots, a_t$ for p such that $(a_1) < (a_1, a_2) < \cdots < p$ is a chain of primes?

5. The ground field k. How does the algebraic closure of k enter into the above discussion? Suppose for the moment that k is an arbitrary field and I is an ideal of $k[X] = k[X_1, \ldots, X_n]$. If $m \supset I$ is a maximal ideal of $k[X]$ for which the coordinates of the "point" (α) determined by $k[X]/m = k[\alpha]$ are in k, then by writing the polynomials of $k[X]$ as polynomials in $k[X-\alpha]$ we may translate (α) to (0) and conclude as before that rank $\mathcal{J}(\alpha)=$ the maximal number $\nu=\nu(I;\alpha)$ of elements from I that are linearly independent $\operatorname{mod} m^2$ over $k=k(\alpha)$. An attempt to reduce to this case by replacing k by the field $k(\alpha)$ leaves rank $\mathcal{J}(\alpha)$ unaltered but possibly decreases ν, so equality over $k(\alpha)$ leads only to the inequality rank $\mathcal{J}(\alpha)\leqslant\nu$. Zariski [74] has proved that equality holds whenever $k(\alpha)$ is separable over k, but not in general, as the following example shows:

Let k be a non-perfect field of characteristic p, so that there exists $a\in k$ such that $a^{1/p}=\alpha_1\notin k$; and consider the ideal $I=(Y^2+X^p-a)$ and the maximal ideal $m=(X^p-a,Y)\supset I$ in $k[X,Y]$. Since $k[X,Y]/m=k[\alpha_1,0]$, the point (α) corresponding to m is $(\alpha_1,0)$; and rank $\mathcal{J}(\alpha)=0$. On the other hand, $Y^2+X^p-a\notin m^2$, so $\nu=1$.

Theorem 4.5 shows that for an arbitrary ground field k, $\nu(I;\alpha)\leqslant \operatorname{ht} I_m$ and that $=$ holds if and only if $k[x]_m$ is a regular local ring (where $k[x]=k[X]/I$). Thus, if we make the distinction that the zero $m(\alpha)$ of I should be called *algebraically simple* for I if $k[x]_m$ is regular, or equivalently, if $\nu(I;\alpha)=\operatorname{ht} I_m$, and *geometrically simple* for I if rank $\mathcal{J}(\alpha)=\operatorname{ht} I_m$, then the content of the above remarks is that geometrically simple implies algebraically simple and that the two concepts coincide whenever $k(\alpha)$ is separable over k.

6. Curves as complete intersections. A variety $\mathcal{V}(I)$ will be called a complete intersection (or, more specifically, an ideal-theoretic complete intersection) if its ideal I is a complete intersection ideal. For an illustration of how this terminology will be expanded upon, 4.5 shows that if a point m is simple on $\mathcal{V}(I)$, then $\mathcal{V}(I)$ is "locally a complete intersection at m," in the sense that the ideal I_m is a complete intersection ideal.

By a *curve* C (in $\mathcal{Q}^n$) we shall mean a 1-dim, unmixed variety; the term *space curve* will be reserved for a curve in 3-space. If I_C is the defining ideal of C, then C is a complete intersection if $\operatorname{ht} I_C=\mu(I_C)=n-1$. There are classical examples due to Macaulay of ht 2 prime ideals p in $k[X_1,X_2,X_3]$ such that $\mu(p)$ is arbitrarily large (cf. §25), but these ideals all define curves with singularities; so let us confine our attention to non-singular curves.

It is also a classical result that any non-singular curve C in $\mathcal{A}^n$ is isomorphic to a (non-singular) curve C' in $\mathcal{A}^3$ (cf. [1, p. 43] for the irreducible case and [23, p. 92] or [64, p. 101] for the general case). This is proved by showing that after applying a suitable k-automorphism of $k[X_1, \ldots, X_n]$, one may assume that the canonical injection $k[X_1, X_2, X_3]/(I_C \cap k[X_1, X_2, X_3]) \to k[X_1, \ldots, X_n]/I_C$ is an isomorphism, from which it follows that I_C is generated by any generating set for $I_{C'} = I_C \cap k[X_1, X_2, X_3]$ together with polynomials $X_i - f_i(X_1, X_2, X_3)$, $i = 4, \ldots, n$. Thus, $\mu(I_C) \leqslant \mu(I_{C'}) + n - 3$; so C is a complete intersection if C' is. We shall see later (§30) that whether or not a non-singular curve is a complete intersection depends only on its affine ring and consequently that if C is a complete intersection, then every curve isomorphic to C is also. Furthermore, in §16 we shall show that $\mu(I_{C'}) \leqslant 3$ and hence that $\mu(I_C) \leqslant n$.

A *surface* S in $\mathcal{A}^3$ will be defined to be a 2-dim, unmixed variety in $\mathcal{A}^3$, or, equivalently, a variety defined by a non-zero principal radical ideal (f) of $k[X_1, X_2, X_3]$, the assumption that (f) is a radical ideal being equivalent to f having no repeated factors in its irreducible decomposition. It follows from the Jacobian criterion that a point (α) of S is simple on S if and only if not all of $\partial f/\partial \alpha_1, \partial f/\partial \alpha_2, \partial f/\partial \alpha_3$ are 0. The *tangent plane* $T(\alpha)$ to S at a simple point (α) of S is defined to be the surface given by $(\partial f/\partial \alpha_1(X_1 - \alpha_1) + \partial f/\partial \alpha_2(X_2 - \alpha_2) + \partial f/\partial \alpha_3(X_3 - \alpha_3))$. Two surfaces (f) and (g) through a common point (α) will be called *transversal* at (α) if (α) is simple on both and their tangent planes at (α) are distinct, or equivalently, if

$$\operatorname{rank}\begin{pmatrix} \partial f/\partial \alpha_1 & \partial f/\partial \alpha_2 & \partial f/\partial \alpha_3 \\ \partial g/\partial \alpha_1 & \partial g/\partial \alpha_2 & \partial g/\partial \alpha_3 \end{pmatrix} = 2.$$

Moreover, since by §3 the rank of this matrix is just the number $\nu(I_m)$ of 4.4, where $I = (f, g)$ and $m = m(\alpha)$, it follows from 4.5 that I_m is a *prime* ideal of ht 2 when this rank is 2, and further, since the ideal of C is ht 2 unmixed and contains I for any curve C

in the intersection of these two surfaces and passing through (α), then I_m must equal the ideal of C localized at m.

A curve C in $\mathfrak{C}^3$ with defining ideal I_C will be called a *set-theoretic complete intersection* of surfaces (f) and (g) if $\mathcal{V}(I_C) = \mathcal{V}((f)) \cap \mathcal{V}((g))$ $(= \mathcal{V}((f, g)))$, or, equivalently, if $I = \sqrt{(f, g)}$. Of course, if C is a complete intersection, then C is also a set-theoretic complete intersection; but conversely, what additional condition is needed in order for a set-theoretic complete intersection to be a complete intersection? A sufficient condition is that the two surfaces (f) and (g) which express C as a set-theoretic complete intersection should be transversal at every point of C; for by the last remark of the preceding paragraph, then f, g generate I_C locally at every maximal ideal of $k[X_1, X_2, X_3]$, from whence it follows that f, g generate I_C [10, p. 112]. In addition, by the Jacobian criterion of §3 this condition implies that C is non-singular. Conversely, if $I_C = (f, g)$ and C is non-singular, then again by the Jacobian criterion, (f) and (g) are transversal at every point of C. In summary, a curve C is a set-theoretic complete intersection of two surfaces (f) and (g) which are transversal at every point of C if and only if C is non-singular and is an ideal-theoretic complete intersection of (f) and (g).

Without going into a discussion of intersection multiplicity, we shall add this concluding remark: Suppose C is a set-theoretic complete intersection of (f) and (g). Then (f) and (g) are transversal at every point of C if and only if C is non-singular and every irreducible component of C occurs with multiplicity 1 in this intersection [73, p. 152].

7. An example. Let C be the variety in $\mathfrak{C}^3$ whose defining ideal I is generated by $f_1 = X_1X_3 - X_2^2$, $f_2 = X_3 - X_1X_2$, $f_3 = X_2 - X_1^2$. Then $X_2 \equiv X_1^2 \bmod I$ and $X_3 \equiv X_1^3 \bmod I$; so for any polynomial $g \in k[X_1, X_2, X_3]$, $g(X_1, X_2, X_3) \equiv g(X_1, X_1^2, X_1^3) \bmod I$. Thus, $g \in I$ if (and only if) $g(t, t^2, t^3) = 0$, where t is a fixed indeterminate. The tuple (t, t^2, t^3) is called a generic point for I, and what we have just verified is that I is the kernel of the homomorphism $k[X_1, X_2, X_3] \to k[t]$ defined by mapping X_i to t^i. Thus, I is a ht 2 prime ideal and C is an irreducible, non-singular space curve with affine ring $k[t]$.

Since $f_1 = X_1 f_2 - X_2 f_3$, we have $I = (f_2, f_3)$. Moreover, one sees easily from the equations $X_2 f_3 = X_1 f_2 - f_1$, $X_3 f_3 = X_2 f_2 - X_1 f_1$ that $I \cap (X_2, X_3) = (f_1, f_2)$ and $I \cap (X_2, X_1) = (f_1, f_3)$; thus the surfaces defined by f_1 and f_2 intersect in $C + L_1$, where L_1 is the X_1-axis, and the surfaces defined by f_1 and f_3 intersect in $C + L_3$, where L_3 is the X_3-axis. The surface $f_1 = 0$ is a cone with vertex at (0), and the idea of studying a space curve as the intersection of a cone and surfaces for which the residual curve is a finite number of lines on the cone is classical and goes back to Cayley (cf. [51, p. 14] and [41, p. 57]).

Note that f_i may also be described as the 2×2 determinant obtained by deleting the ith column of the matrix

$$\begin{pmatrix} 1 & X_1 & X_2 \\ X_1 & X_2 & X_3 \end{pmatrix}.$$

In other words, I is a "determinantal ideal." This is not an isolated phenomenon: the ideal of any non-singular curve in $\mathcal{Q}^3$ has a generating set consisting of the 2×2 subdeterminants of a 2×3 matrix with entries from $k[X_1, X_2, X_3]$. (This follows from Burch's theorem [32, p. 148] after one has first proved that I has a resolution of the form $0 \to R^2 \to R^3 \to I \to 0$, where $R = k[X_1, X_2, X_3]$.) See §22 for a further discussion.

It is also of interest to consider the corresponding homogeneous ideal $I^h = (f_1^h, f_2^h, f_3^h)$ in $k[X_0, X_1, X_2, X_3]$, where $f_1^h = f_1$, $f_2^h = X_0 X_3 - X_1 X_2$, and $f_3^h = X_0 X_2 - X_1^2$. This ideal may be seen to be prime (cf. [24, p. 206] for a detailed discussion of homogeneous ideals generated by the 2×2 subdeterminants of a 2×3 matrix whose entries are linear forms in $k[X_0, X_1, X_2, X_3]$). Since the three forms f_i^h are linearly independent over k, they constitute a minimal generating set for I^h (cf. §8); and thus the non-singular irreducible *projective* space curve defined by I^h is not an ideal-theoretic complete intersection of two surfaces, as one may also conclude directly from Bezout's theorem [24, p. 176].

8. Homogeneous complete intersection ideals. A variety in projective n-space is defined by a homogeneous radical ideal $I \neq (X_0, X_1, \ldots, X_n)$ in $R = k[X_0, X_1, \ldots, X_n]$, and the corresponding affine variety $\mathcal{V}(I)$ in $\mathcal{Q}^{n+1}$ is a cone with vertex at the origin. An advantage in working with a homogeneous ideal I is that $\mu(I) = \mu(I_\theta)$, where $\theta = (X_0, \ldots, X_n)$. For if $f_1, \ldots, f_t$ is a set of forms (i.e., homogeneous polynomials) which generate I, then a minimal generating set $f'_1, \ldots, f'_s$ for I_θ may be selected from among these; and $f'_1, \ldots, f'_s$ must also (minimally) generate I, as one sees by writing $gf_i = \sum_j h_j f'_j$, $g, h_j \in R$, $g \notin \theta$, and equating forms of least degree. Thus, complete intersection questions for homogeneous ideals are local.

On the other hand, even if the projective variety one begins with is non-singular, the corresponding affine cone, with the exception of a linear variety, has a singularity at its vertex θ; algebraically, the local ring R_θ / I_θ is not regular even though every localization of it at a non-maximal prime is regular. This is sometimes expressed by saying that R_θ / I_θ has an "isolated singularity." It is in this setting that Hartshorne in his survey article [25, §5] discusses conditions for a homogeneous prime I with $\operatorname{ht} I$ small and n large to be a c.i. ideal. On the negative side, Geyer [23, p. 100] has shown by elementary methods that the ideal of a non-singular irreducible curve in projective 3-space may require an arbitrarily large number of generators.

A related kind of problem has been studied by Peskine–Szpiro in [57, §7]. They give conditions under which the following implication holds: If I is a $\operatorname{ht} t$ homogeneous ideal of $k[X_0, \ldots, X_n]$ such that $(X_0, \ldots, X_n)$ is not an associated prime of I, then $\mu(I) > t + 1$ implies $\mu(I \cap q) > t + 1$ for every ideal q such that $\sqrt{q} = (X_0, \ldots, X_n)$.

9. Ideal-theoretic versus set-theoretic. We shall survey here the possibilities for a non-singular irreducible space curve C. First consider an affine curve. Abhyankar [1, p. 43] and Murthy [48] have proved independently that $\mu(I_C) \leqslant 3$ (cf. §16), and they have given examples showing that this bound is the best possible (see also [3]). Moreover, Ferrand and Szpiro have recently proved (cf. [49]) that C is always a set-theoretic complete intersection.

On the other hand, in the case of a projective curve C it becomes more difficult to cut out C completely in the set-theoretic sense, since one no longer has the option of arranging for a residual intersection to be lost in the hyperplane at infinity; thus, it is still an open question whether such a curve is a set-theoretic complete intersection. It is known, though, by a theorem of M. Kneser [36] that *any* projective space curve may be cut out completely in the set-theoretic sense by three surfaces (cf. [19], [69] for generalizations to n-space). On the other hand, the ideal-theoretic question for a projective C may be quickly disposed of since, as noted in §8, Geyer has proved that the ideal for C may require an arbitrarily large number of generators.

In addition to (i) set-theoretically and (ii) ideal-theoretically, there is yet a third way in which a non-singular irreducible space curve C may be regarded as an intersection of surfaces, namely, (iii) set-theoretically with multiplicity 1; and, as was noted in the last remark of §6, (ii) and (iii) are equivalent so far as the possibility of expressing an affine C as a complete intersection (of two surfaces) is concerned. However, for three surfaces it is no longer clear just what the added requirement "with multiplicity 1" means, and there is an interesting history connected with this. In 1891 Vahlen, in response to a question of Kronecker, gave an example of a non-singular irreducible projective space curve which is not a complete intersection of three surfaces; but it was not made clear whether this was intended in the sense of (i) or (iii), and the resulting confusion led to a fascinating exchange between the two venerable mathematicians, Perron and Severi. In 1941 Perron discovered that there exist three surfaces which cut out the Vahlen example completely in the set-theoretic sense, but in a second paper he mentions that some (unnamed) readers had accused him of misunderstanding the intent of the original Kronecker question and Vahlen's example. Shortly thereafter a paper of Severi appeared in which he revealed himself to be one of these critics, claiming that the Vahlen example should be interpreted in the sense of (iii), thereby providing a legitimate counterexample to Kronecker's question. Perron responded with a careful analysis of Severi's paper and concluded that Severi himself did not know precisely what it means for *three* surfaces to intersect in a curve with multiplicity 1, to which Severi later replied with a more detailed explanation. Even though the argument degenerates at one point to a misunderstanding over use of the word "curve," with Severi arguing that unless expressly stated to the contrary "curve" traditionally had meant "nonsingular curve," the dispute nonetheless led to much of the present-day interest in set-theoretic complete intersections (cf. [68] and [56]).

II. A LOCAL-GLOBAL PRINCIPLE

Let $I \neq R$ be a finitely generated regular ideal of a ring R. The Forster–Swan theorem (§14) tells one how to use a collection of local bounds for $\{\mu(I_p)\}$ to obtain a global bound for $\mu(I)$. Serre observed that if one restricts attention to an ideal having a resolution of the form $0 \to R^t \to P \to I \to 0$ with P a f.g. projective R-module (e.g., any ht 2 unmixed ideal of $k[X, Y, Z]$; cf. §17), then one can improve on the Forster-Swan estimate by the following procedure. Since $\operatorname{rk} I = 1$, $\operatorname{rk} P = t + 1$. Therefore if $\mu_1(I)$ is defined to be the least such t and if all f.g. R-projectives are free, then $\mu(I) = \mu_1(I) + 1$. Thus, an upper bound for $\mu(I)$ will lead to one for $\mu_1(I)$, and, conversely, provided f.g. R-projectives are free. In particular, since projectives are free over a local ring, one can translate local data $\{\mu(I_p)\}$ to local data $\{\mu_1(I_p)\}$ by means of this equality. At this point Serre's characterization $\mu_1(I) = \mu(\operatorname{Ext}^1_R(I, R))$ becomes pertinent; for it allows one to apply the Forster-Swan theorem to the R-module $\operatorname{Ext}^1_R(I, R)$, thereby carrying down a bound for $\{\mu_1(I_p)\}$ to one for $\mu_1(I)$ and hence ultimately to one for $\mu(I)$. The advantage in applying the Forster-Swan theorem to $\operatorname{Ext}^1_R(I, R)$ rather than directly to I lies in the fact that, in order to apply the theorem to an R-module N, one need only know local data at the collection of primes containing $\operatorname{Ann} N$; if $N = I$, this collection consists of all primes of R, whereas if $N = \mathrm{Ext}_R^1(I, R)$, it consists only of primes containing I.

We shall work out some of the details of this local-global principle in this section and shall show how it can be used to prove that the ideal of a non-singular space curve is generated by $\leqslant 3$ elements.

Fix throughout section II a ring R and a f.g. R-module M such that $d(M) = 1$. Here $d(M)$ denotes f.g. projective dimension; it will suffice for the present to know that $d(M) = 1$ means there exists an exact sequence of R-modules $0 \to P_1 \to P_0 \to M \to 0$ with P_0, P_1 f.g. projective but M itself not projective.

10. The theorem of Serre-Murthy. The central theorem of this section is the following.

10.1. THEOREM (Serre [67], Murthy [48]): *Suppose $d(M) = 1$. Then the following positive integers are equal:*

(i) $\inf\{t \mid$ *there exists an exact sequence* $0 \to R^t \to P \to M \to 0$ *with* P f.g. *projective*$\}$.

(ii) $\inf\{\mu(P_1) \mid$ *there exists an exact sequence* $0 \to P_1 \to P_0 \to M \to 0$ *with* P_0, P_1 f.g. *projective*$\}$.

(iii) $\mu(\mathrm{Ext}_R^1(M, R))$.

Notation: Let $\mu_1(M)$ denote the positive integer defined by (i); in this notation the equality of (i) and (iii) reads $\mu_1(M) = \mu(\mathrm{Ext}_R^1(M, R))$.

Note first that (i) = (ii) is immediate. For, given any exact sequence $0 \to P_1 \to P_0 \to M \to 0$ with P_0, P_1 f.g. projective, there exists an exact sequence $0 \to Q \to R^t \to P_1 \to 0$ with $\mu(P_1) = t$. Since P_1 is projective, $Q \oplus P_1 \cong R^t$; and therefore by adding Q to the terms P_1, P_0 of the given sequence, one obtains an exact sequence $0 \to R^t \to P_0 \oplus Q \to M \to 0$, where $P_0 \oplus Q$ is again f.g. projective since P_0 and Q are.

Before turning to the proof of (i) = (iii), let us review some facts about Ext_R^1 and f.g. projective dimension.

11. Ext_R^1. (Cf. [44, Chapter III].) An exact sequence $E: 0 \to A \to B \to C \to 0$ of R-modules is called an extension of A by C. Two such extensions of A by C are defined to be equivalent if they differ by an isomorphism in the middle term which induces equality on the end terms. The set of equivalence classes of such extensions can be given an R-module structure, and the resulting R-module is denoted $\mathrm{Ext}_R^1(C, A)$.

In particular, scalar multiplication is defined as follows. For $r \in R$, use r to also denote the multiplication homomorphism of $A \to A$ defined by $r(a) = ra$. Then given any extension E, there exists an extension E', unique up to equivalence, for which there exists a homomorphism $B \to B'$ making the following diagram commutative:

$$\begin{array}{lccccccccc} E': & 0 & \to & A & \to & B' & \to & C & \to & 0 \\ & & & r\uparrow & & \uparrow & & \| & & \\ E: & 0 & \to & A & \to & B & \to & C & \to & 0; \end{array}$$

and $r[E]$ is defined to be $[E']$. (Here $[E]$ denotes the equivalence class of E in $\mathrm{Ext}_R^1(C, A)$.)

Properties: (a) $\mathrm{Ext}_R^1(C, A) = 0$ if and only if every extension of A by C splits. In particular, if C is projective, then $\mathrm{Ext}_R^1(C, A) = 0$.

(b) Given the exact sequence E above and an R-module N, there exists an exact sequence of R-modules

$$\begin{aligned} &0 \to \mathrm{Hom}_R(C, N) \to \mathrm{Hom}_R(B, N) \to \\ &\mathrm{Hom}_R(A, N) \xrightarrow{\phi} \mathrm{Ext}_R^1(C, N) \to \mathrm{Ext}_R^1(B, N) \to \\ &\mathrm{Ext}_R^1(A, N) \to \mathrm{Ext}_R^2(C, N) \to \cdots. \end{aligned}$$

Moreover, if $A = R$ in E, so that E has the form $0 \to R \to B \to C \to 0$, and if $N = R$ also, then the connecting homomorphism ϕ is defined by $\phi(1) = [E]$, where 1 denotes the identity homomorphism in $\mathrm{Hom}_R(R, R)$. If in addition B is projective, so that $\mathrm{Ext}_R^1(B, R) = 0$ and hence ϕ is surjective, then $[E]$ generates $\mathrm{Ext}_R^1(C, R)$.

(c) By applying (b) and (a) to a short exact sequence $0 \to K \to P \to C \to 0$ with P projective, one obtains an exact sequence

$$0 \to \operatorname{Hom}_R(C,N) \to \operatorname{Hom}_R(P,N) \to \operatorname{Hom}_R(K,N)$$

$$\to \operatorname{Ext}^1_R(C,N) \to 0.$$

By localizing at a given m.s. S of R both before and after performing this operation and using the fact that $\operatorname{Hom}_R(L,N)_S$ is canonically isomorphic to $\operatorname{Hom}_{R_S}(L_S,N_S)$ whenever L is finitely presented [32, p. 163], one sees that $\operatorname{Ext}^1_R(C,N)_S \cong \operatorname{Ext}^1_{R_S}(C_S,N_S)$ whenever K may be chosen finitely presented and P finitely generated (e.g., if $d(C)=1$). Moreover, if $d(C)=1$, then K may even be chosen f.g. free, in which case $\operatorname{Hom}_R(K,R)$, and hence also $\operatorname{Ext}^1_R(C,R)$, is finitely generated.

(d) If I is an ideal of R such that $d(I)=1$, then $V(\operatorname{Ann}\operatorname{Ext}^1_R(I,R)) \subset V(I)$. For if p is a prime of R, then $p \not\supset I \Rightarrow I_p = R_p \Rightarrow \operatorname{Ext}^1_{R_p}(I_p,R_p)=0 \Rightarrow \operatorname{Ext}^1_R(I,R)_p = 0 \Rightarrow p \not\supset \operatorname{Ann}\operatorname{Ext}^1_R(I,R)$, the second implication by (a), the third by (c), and the fourth by the observation of (c) that $\operatorname{Ext}^1_R(I,R)$ is finitely generated.

(e) $\operatorname{Ext}^1_R(C,A_1 \oplus A_2) \cong \operatorname{Ext}^1_R(C,A_1) \oplus \operatorname{Ext}^1_R(C,A_2)$ for R-modules C, A_1, A_2. In particular, if C has a projective resolution of the form $0 \to R^t \to P \to C \to 0$, then C is projective $\Leftrightarrow \operatorname{Ext}^1_R(C,R^t) = 0 \Leftrightarrow \operatorname{Ext}^1_R(C,R)=0$. Thus, $\operatorname{Ext}^1_R(C,R) \neq 0$ if $d(C)=1$.

12. Finitely generated projective dimension. (Cf. [54].) Let N be an R-module. An exact sequence $0 \to P_n \to P_{n-1} \to \cdots \to P_0 \to N \to 0$ with the P_i f.g. projective R-modules is called a f.g. projective resolution of length n for N. A f.g. projective resolution of length ∞ is defined similarly. For an N having such a resolution (of length $\leqslant \infty$), the f.g. projective dimension $d(N)$ is said to be defined and is given by $d(N) = \inf\{n \mid N$ has a f.g. projective resolution of length $n\}$. The main property of $d(N)$ is summed up by the dimension theorem: *If* $0 \to A \to B \to C \to 0$ *is exact and d is defined for any two of* A, B, C, *then it is defined for the third*; *and when this is so, then* $d(B) \leqslant \max\{d(A), d(C)\}$, *and the inequality implies* $d(C) = d(A)+1$. It follows from this, for example, that if $d(N) \leqslant n$ and $0 \to K \to P_{n-1} \to \cdots \to P_0 \to N \to 0$ is exact with the P_i f.g. projective, then K is also f.g. projective.

Moreover, if $d(N)$ is defined, then $d(N) = \sup\{d(N_p) \mid p \in \operatorname{Spec} R\}$; for the nth kernel K in a f.g. projective resolution for N is automatically finitely presented, and a f.g. module K is projective if and only if K is finitely presented and K_p is projective for every $p \in \operatorname{Spec} R$ [10, p. 138].

Finally, note that f.g. projective dimension coincides with the usual notion of projective dimension for a f.g. module N over a *noetherian* ring R.

13. Proof of (i) = (iii) of the Serre-Murthy theorem. First observe that if there exists an exact sequence $E: 0 \to R^t \to P \to M \to 0$ with P f.g. projective, then $\mu(\operatorname{Ext}^1_R(M,R)) \leqslant t$. For by 11(a), (b), E yields an exact sequence $\cdots \to \operatorname{Hom}_R(R^t, R) \to \operatorname{Ext}^1_R(M, R) \to 0$; and since $\operatorname{Hom}_R(R^t, R)$ is free on t generators, $\operatorname{Ext}^1_R(M,R)$ is then generated by t elements.

Thus, it remains to prove that, conversely, if $\mu(\operatorname{Ext}^1_R(M,R)) \leqslant t$, then there exists an exact sequence of the form $0 \to R^s \to Q \to M \to 0$ with $s \leqslant t$ and Q f.g. projective. Note that $t \geqslant 1$ since M is not projective implies $\operatorname{Ext}^1_R(M,R) \neq 0$ by 11(e). Let $[E_1], \ldots, [E_t]$ be a generating set for $\operatorname{Ext}^1_R(M,R)$, and let E_1 be the exact sequence $0 \to R \to L \to M \to 0$. By 11(b), there exists an exact sequence $\cdots \to \operatorname{Hom}_R(R,R) \xrightarrow{\phi} \operatorname{Ext}^1_R(M,R) \xrightarrow{\psi} \operatorname{Ext}^1_R(L,R) \to 0$, where ϕ maps the identity homomorphism to $[E_1]$. Since $[E_1]$ is in the image of ϕ, $\psi([E_1]) = 0$; and therefore either $t = 1$ and $\operatorname{Ext}^1_R(L,R) = 0$, or $t > 1$ and $\operatorname{Ext}^1_R(L,R)$ is generated by the $t-1$ elements $\psi([E_2]), \ldots, \psi([E_t])$.

Continue by induction on t. By §12, $d(M) = 1$ implies $d(L) \leqslant 1$; and if $d(L) = 0$, then L is f.g. projective and E_1 is the required sequence. In particular, if $t = 1$, then $\operatorname{Ext}^1_R(L,R) = 0$ implies L is projective by 11(e); and hence E_1 is the required sequence. Therefore we may assume $t > 1$, $d(L) = 1$, and $\operatorname{Ext}^1_R(L,R)$ is generated by $t-1$ elements. By induction hypothesis there exists an exact sequence $0 \to R^s \to Q \to L \to 0$ with Q f.g. projective and $s \leqslant t-1$. Now it is easily seen that there exists an R-module K which gives rise to the following diagram with exact rows and columns (K is a "pull back"; see, for example, [54]):

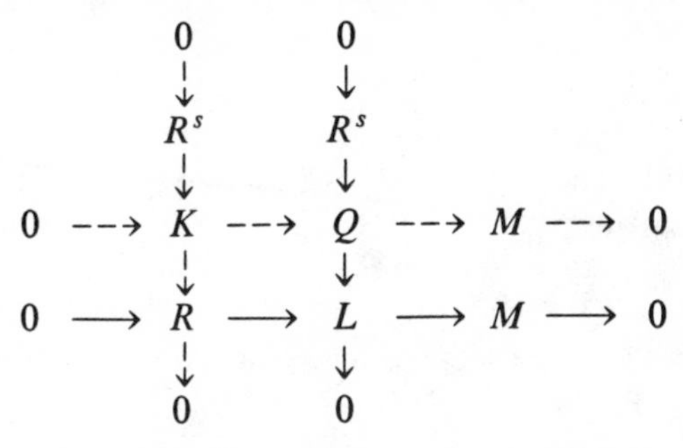

Exactness of the left column shows $K \cong R^s \oplus R$, and then the upper row is the exact sequence needed to conclude the proof. Q.E.D.

13.1. COROLLARY: *Let I be a regular ideal of a ring R, and suppose $d(I) = 1$ and* f.g. *R-projectives are free. Then $\mu(I) = \mu(\operatorname{Ext}^1_R(I, R)) + 1$.*

It is good to keep in mind the case of a polynomial ring over a field; and if one glances ahead to §17, he will see that the hypothesis of the following corollary implies $d(I) = 1$.

13.2. COROLLARY: *Suppose I is a* ht 2 *unmixed ideal of $R = k[X_1, \ldots, X_n]$, k a field. If I is locally a* c.i. *ideal at every prime containing I, then I is a* c.i. *ideal if and only if $\operatorname{Ext}^1_R(I, R)$ is cyclic.*

Theorem 10.1 generalizes in a rather superficial way to any M such that $d(M) < \infty$, but the generalization does not seem to be very useful:

13.3. GENERALIZATION OF THE SERRE-MURTHY THEOREM: *Suppose $d(M) = n \geqslant 1$. Then the following positive integers are equal:*

(i) $\inf\{t \mid$ *there exists an exact sequence* $0 \to R^t \to P_{n-1} \to \cdots \to P_0 \to M \to 0$ *with the P_i* f.g. *projective*$\}$.

(ii) $\inf\{\mu(P_n) \mid$ *there exists an exact sequence* $0 \to P_n \to P_{n-1} \to \cdots \to P_0 \to M \to 0$ *with the* P_i f.g. *projective*$\}$.
(iii) $\mu(\mathrm{Ext}_R^n(M,R))$.

Proof: The proof of (i) = (ii) is the same as in the proof of theorem 10.1. The proof of (i) = (iii) is by induction on n, the $n=1$ case being theorem 10.1. For $n>1$, choose a resolution $0 \to R^t \to P_{n-1} \to \cdots \to P_1 \to P_0 \to M \to 0$ with t minimal. If $K = \ker(P_1 \to P_0)$, then by induction hypothesis $\mu(\mathrm{Ext}_R^{n-1}(K,R)) = t$. But by the long exact sequence for Ext (cf. §11(b)) applied to $0 \to K \to P_0 \to M \to 0$, $\mathrm{Ext}_R^{n-1}(K,R) \cong \mathrm{Ext}_R^n(M,R)$. Q.E.D.

14. The Forster-Swan theorem. (Cf. [34, p. 25].) Let N be a f.g. R-module. We now want to find a bound for $\mu_R(N)$ whenever a bound for $\{\mu_{R_p}(N_p)\}$ is given. Roughly speaking, the Forster-Swan theorem asserts that if $\mu(N_p) \leqslant \beta$ for every $p \in \mathrm{Spec}\, R$ and $\alpha = \mathrm{coht}(\mathrm{Ann}\, N)$, then $\mu(N) \leqslant \beta + \alpha$.

To state the general form of the theorem, we need some additional terminology. A *j-radical* ideal of R is an ideal which is an intersection of maximal ideals; if p is a prime j-radical ideal, we shall call it a *j-prime* for short. The ring R is said to have *noetherian j-spectrum* if R has ascending chain condition on j-radical ideals. The j-dim of R is the maximal length of a chain of j-primes, and the j-coht of an ideal I is the maximum length of a chain of j-primes ascending from I. Furthermore, just as $V(I)$ was used for the set of primes $\supset I$, we shall use $V_j(I)$ to denote the set of j-primes $\supset I$.

The letter j may be dropped throughout this discussion without the results losing significance, although their applicability will usually be considerably diminished. For instance, a quasi-local ring is always j-noetherian of j-dim 0, even though there exist quasi-local rings of infinite Krull dim. On the other hand, there is a large class of rings, among them the affine rings, for which the j-primes and the primes coincide; these are called Jacobson-Hilbert rings. Similarly, for a polynomial ring $S[X]$, whether one uses primes or j-primes is often irrelevant, mainly because every extended prime of $S[X]$ is a j-prime; thus, $S[X]$ has noetherian

j-spectrum if and only if $S[X]$ has noetherian spectrum, and $j\text{-dim}\, S[X] = \dim S[X]$. Note, however, that if S has noetherian j-spectrum, it does not follow that $S[X]$ does also; cf. [55].

For any j-prime p of R, define the "corrected" mininal number of generators for N at p by

$$\mu^*(p, N) = \mu(N_p) + j\text{-coht}\, p.$$

FORSTER-SWAN THEOREM: *Suppose the ring R has noetherian j-spectrum and is of finite j-*dim, *and let $N \neq 0$ be a* f.g. *R-module. Then* $\mu(N) \leqslant \sup\{\mu^*(p, N) \mid p \in V_j(\operatorname{Ann} N)\}$.

For example, if R is a Dedekind domain with infinitely many maximal ideals and $I \neq 0$ an ideal of R, then $\mu^*(p, I) = 1$ for every maximal prime p and $\mu^*((0), I) = 2$; so the theorem asserts that $\mu(I) \leqslant 2$.

Before proceeding to the application we have in mind, it is necessary to develop an appropriate concept of rk in order to relate $\mu(M)$ and $\mu_1(M)$.

15. Rank of a module. For a f.g. R-module N, we shall define $\operatorname{rk} N$ to be $\inf\{\mu(N_p) \mid p \text{ is a ht}\,0 \text{ prime}\}$. Note that since $p \subset q$ implies $\mu(N_p) \leqslant \mu(N_q)$, we also have $\operatorname{rk} N = \inf\{\mu(N_p) \mid p \text{ is a prime ideal}\}$. For example, if R is a domain, then $\operatorname{rk} N = \mu(N_{(0)})$; or if $N = R^t$, then $\operatorname{rk} N = t$. N will be said to have (locally) constant $\operatorname{rk} t$ on a set of primes $\mathbb{S}$ if $t = \operatorname{rk} N_p$ for all $p \in \mathbb{S}$, or, equivalently, if $\mu(N_p) = t$ for every ht 0 prime p which is contained in a prime of $\mathbb{S}$; when $\mathbb{S} = \operatorname{Spec} R$, we omit explicit reference to $\mathbb{S}$.

Given a short exact sequence $0 \to A \to B \to C \to 0$ of f.g. R-modules, under what conditions does $\operatorname{rk} B = \operatorname{rk} A + \operatorname{rk} C$? This is certainly the case if R is a domain, but in general the additivity may fail even if the sequence splits. However, it does hold if the sequence is of the form $0 \to R^t \to P \to M \to 0$ with P f.g. projective, as one sees from the following two facts: (i) For such an M, M_p is free for any ht 0 prime p [32, p. 139], and hence the sequence splits at such a p and $t + \mu(M_p) = \mu(P_p)$. (ii) Since $t = \operatorname{rk} R^t$ is constant, $\mu(P_p)$ and $\mu(M_p)$ necessarily attain their infs at the same ht 0 prime

p. If, in addition, R is a ring for which f.g. projectives are free (more precisely, f.g. projectives of rk equal to $\mu_1(M) + \operatorname{rk} M$), then by choosing a sequence such that $t = \mu_1(M)$, one finds that $\mu(M) \leqslant \mu(P) = \operatorname{rk} P = \mu_1(M) + \operatorname{rk} M$. By similar reasoning applied to a sequence of the form $0 \to Q \to R^s \to M \to 0$ with $s = \mu(M)$ and Q f.g. projective, one finds that $\operatorname{rk} Q + \operatorname{rk} M \leqslant \mu(M)$; and if Q is free, then $\operatorname{rk} Q = \mu(Q) \geqslant \mu_1(M)$ and hence $\mu_1(M) + \operatorname{rk} M \leqslant \mu(M)$. In summary, if $d(M) = 1$ and f.g. R-projectives are free, then $\mu(M) = \mu_1(M) + \operatorname{rk} M$.

We record in the next lemma one further fact for later use. Since $\mu(N)$ does not increase under localization, $\operatorname{rk} N \leqslant \mu(N)$ is always valid.

15.1. LEMMA: *If R is a reduced ring and N is a* f.g. *R-module, then* $\operatorname{rk} N = \mu(N)$ *(if and) only if N is free.*

Proof: Suppose $\operatorname{rk} N = \mu(N) = t$. Then $\mu(N_p) = t$ for every $p \in \operatorname{Spec} R$. Let $x_1, \ldots, x_t$ be a minimal generating set for N, and suppose $\sum_{i=1}^{t} r_i x_i = 0$, $r_i \in R$. Since this relation also holds locally and $\mu(N_p) = t$, no r_i localizes to a unit in R_p. Therefore, $r_i \in \cap \{p \mid p \in \operatorname{Spec} R\} = 0$, $i = 1, \ldots, t$. Q.E.D.

16. Application to a bound for $\mu(M)$.

16.1. THEOREM: *Let R be a ring having noetherian j-spectrum and finite* j-dim, *let M be a* f.g. *R-module such that $d(M) = 1$, and let* $A_1 = \operatorname{Ann} \operatorname{Ext}_R^1(M, R)$. *Then* $\mu_1(M) \leqslant \sup\{\mu^*(p, M) \mid p \in V_j(A_1)\} - \operatorname{rk} M$; *and if, in addition,* f.g. *projective R-modules are free, then* $\mu(M) \leqslant \sup\{\mu^*(p, M) \mid p \in V_j(A_1)\}$.

Proof: Since projectives are free over a quasi-local ring, by §15 we have $\mu(M_p) = \mu_1(M_p) + \operatorname{rk} M_p$ for any prime $p \in V_j(A_1)$. Then $\mu^*(p, M) = \mu(M_p) + j\text{-coht } p = \mu_1(M_p) + \operatorname{rk} M_p + j\text{-coht } p = \mu(\operatorname{Ext}_R^1(M_p, R_p)) + \operatorname{rk} M_p + j\text{-coht } p = \mu(\operatorname{Ext}_R^1(M, R)_p) + j\text{-coht } p + \operatorname{rk} M_p = \mu^*(p, \operatorname{Ext}_R^1(M, R)) + \operatorname{rk} M_p \geqslant \mu^*(p, \operatorname{Ext}_R^1(M, R)) + \operatorname{rk} M$. By the Forster-Swan theorem applied to the local bounds $\mu^*(p, M) - \operatorname{rk} M \geqslant \mu^*(p, \operatorname{Ext}_R^1(M, R))$, we then have $\mu_1(M)$

$= \mu(\mathrm{Ext}_R^1(M,R)) \leqslant \sup\{\mu^*(p,M) \mid p \in V_j(A_1)\} - \mathrm{rk}\, M$. For the second assertion of the theorem, note that by §15, $\mu(M) = \mu_1(M) + \mathrm{rk}\, M$ if f.g. R-projectives are free. Q.E.D.

REMARK: The theorem may be sharpened slightly as follows: Define $\mathrm{rk}_{A_1} M$ to be $\inf\{\mu(M_p) \mid \mathrm{ht}\, p = 0 \text{ and } (p, A_1) \neq R\}$. Then replace $\mathrm{rk}\, M$ by $\mathrm{rk}_{A_1} M$; and for the second assertion, specify only that f.g. projectives P such that $\mathrm{rk}_{A_1} P = \sup\{\mu^*(p,M) \mid p \in V_j(A_1)\}$ be free.

The crux of this theorem is that by strengthening the hypothesis of the Forster-Swan theorem, one can improve the conclusion in that one only need consider the j-primes containing Ann $\mathrm{Ext}_R^1(M,R)$, rather than the possibly larger set of j-primes containing Ann M. For example, consider the case of the theorem where $M = I$ is a non-zero ideal of $R = k[X_1, \ldots, X_n]$. The Forster-Swan theorem requires a bound for $\mu^*(p,I)$ for every j-prime p containing $0 = \mathrm{Ann}\, I$, and hence for *every* j-prime p. On the other hand, the above theorem only requires looking at those j-primes p containing Ann $\mathrm{Ext}_R^1(I,R)$, and by 11(d), every such p contains I.

Since our main concern will be with $M =$ a regular ideal I, we shall rephrase the theorem for this case.

16.2. COROLLARY: *Let R be a ring having noetherian j-spectrum and finite j-*dim, *and let I be a regular ideal of R such that $d(I) = 1$. If $\beta = \sup\{\mu^*(p,I) \mid p \in V_j(I)\}$, then there exists an exact sequence $0 \to R^{\beta-1} \to P \to I \to 0$ with P* f.g. *projective (of constant* rk β). *If, in addition,* f.g. *R-projectives of constant* rk β *are free (or, more specifically, if P is free), then $\mu(I) \leqslant \beta$.*

For another comparison of the bound on $\mu(I)$ given by this corollary with that of the Forster-Swan theorem, one may think of the Forster-Swan bound as being $\mu(I) \leqslant \sup\{\mu^*(p,I), d+1 \mid p \in V_j(I)\}$, where $d = j\text{-}\dim R$, whereas the corollary yields the bound $\mu(I) \leqslant \sup\{\mu^*(p,I) \mid p \in V_j(I)\}$.

Consider the case of a ht 2 unmixed ideal I of $R = k[X,Y,Z]$, k a field. We shall show in section III that $d(I) = 1$; and by the

Quillen and Suslin theorem it is known that f.g. R-projectives are free. (To apply the corollary, one need only know that f.g. projectives of rk 3 are free.) Therefore if I_p is generated by 2 elements at every prime $p \supset I$ (e.g., if R/I is regular; cf. 4.6), it follows from the corollary that $\mu(I) \leqslant 3$. This result was established independently by Abhyankar [1, p. 43] and Murthy [48]. The approach described above is Murthy's; and, of course, the main difficulty lies in verifying that f.g. rk 3 projectives over $k[X,Y,Z]$ are free. Another way of thinking of this example is that one is allowed to ignore the primes of coht 2 and 3 of $k[X,Y,Z]$ in computing the Forster-Swan bound for $\mu^*(p,I)$; note that one would get the same bound by merely ignoring the coht 3 prime (0). Eisenbud-Evans [18] have ventured a general conjecture along these lines, which asserts that for any polynomial ring $S[X]$, S noetherian of finite dim, one should be able to ignore the primes of maximal coht in computing the Forster-Swan bound.

REMARKS: (a) The ideas of section II have their source in the fundamental papers [66], [67] of Serre. Theorem 10.1 was first proved by Serre for the case that $\mu_1(M)=1$ and later generalized by Murthy [48]. Actually, the original versions involved assumptions that R be noetherian and that Spec R be connected, and we have eliminated these in Theorem 10.1 and §16 by merely using f.g. projective dim instead of projective dim and by using an appropriate notion of rk. Of course, Serre's work preceded that of Forster and Swan, and the point of view that we take in regarding the theorem of §16 as a kind of Forster-Swan theorem presents an admittedly inaccurate historical perspective.

(b) Eisenbud-Evans [17] have put the Forster-Swan theorem in a very general setting, thereby clarifying its relationship with a number of other theorems. See also [12] for improvements on the Eisenbud-Evans work.

III. WHEN IS $d(I)=1$?

Throughout section III, R will denote a *noetherian* ring, unless explicitly stated otherwise, and I an ideal of R different from R.

17. Statement of the main theorem. In order to apply the local-global principle of section II to an ideal I, it is first necessary to know whether $d(I)=1$. We shall collect in the next theorem some rather general criteria for this to happen. For example, part (a) includes the fact that any ht 2 unmixed ideal of $k[X_1,\ldots,X_n]$ which is locally a complete intersection satisfies $d(I)=1$, and part (b) applied to $k[X,Y,Z]$ shows that the ideals of $k[X,Y,Z]$ with $d(I)=1$ are exactly the non-zero constant multiples of ht 2 unmixed ideals.

17.1. THEOREM: *Suppose* $d(I)<\infty$.

(a) *If* $\sup\{\mu(I_p) \mid p \in V(I)\}=2$, *then* $d(I)=1$.

(b) *If I is a regular ideal such that* $d(I)=1$, *then* $I=I_0I^*$, *where* I_0 *is an invertible ideal and* I^* *is a proper* ht 2, *grade* 2, *grade-unmixed ideal; and the converse holds if R is* CM *and all maximal ideals have* ht 3.

Before turning to the proof, we shall try to place the theorem in perspective by looking at some of the algebra surrounding it. For the moment, note that there is no hope for a converse to (a), as the Macaulay examples demonstrate. Also, in (b) the invertible factor I_0 is to be expected; for multiplying an ideal I by a regular principal ideal (a), or, more generally, by an invertible ideal, which is locally of this type, does not alter $d(I)$. On the other hand, if $(a)\neq R$, $\operatorname{ht}(a)I \leqslant 1$, even though $\operatorname{ht} I$ may be large. (Actually, $d(I)=1$ implies $\operatorname{ht} I \leqslant 2$, which is a known special case of the homological height conjecture; cf. [27, p. 143].)

18. Invertible ideals. An ideal I of an arbitrary ring R will be called *invertible* if I is a f.g. regular ideal which is locally principal. Recall that for a f.g. R-module N, N is flat if and only if N is locally free, and N is projective if and only if N is locally free and has constant rk locally on a neighborhood of every prime [10, p. 138]; so other ways of describing an invertible ideal I are that I should be a f.g. flat regular ideal or that I should be a f.g. projective regular ideal of constant rk 1. With respect to this latter description, we have

18.1. PROPOSITION: *Let R be a noetherian ring. Then any* f.g. *projective R-module P of constant* rk 1 *is isomorphic to an invertible ideal of R.*

Proof: If S is the set of regular elements of R, then P_S is R_S-free. (More generally, if P is a f.g. projective R-module of constant rk and R is a ring, not necessarily noetherian, with only finitely many maximal ideals, then P is free: reduce to the case that R is a finite product of fields by passing to $R/j(R)$, where $j(R)$ is the intersection of the maximal ideals of R [10, p. 143].) Therefore, $P_S \cong R_S$ by the rk 1 assumption. But the canonical homomorphism $P \to P_S$ is injective since P is torsion-free; so P is (isomorphic to) a submodule of R_S. Choosing a common denominator, we then get an $s \in S$ such that $sP \subset R$; and hence sP is an ideal of R. But s is regular implies $P \cong sP$.

To complete the proof, it remains to observe that an ideal I of a noetherian ring R which is locally free of constant rk 1 is regular. For $0 = \operatorname{Ann}_{R_p} I_p = (\operatorname{Ann}_R I)_p$ for every prime p implies $\operatorname{Ann}_R I = 0$; and since R is noetherian, this implies $I \not\subset \mathfrak{Z}(R)$. Q.E.D.

REMARK: For an arbitrary ring R, a projective R-module P of constant rk 1 is necessarily f.g. [71, p. 431], but I do not know if such a P need be isomorphic to an ideal of R. In any event, a projective ideal of constant rk 1 need not be regular if R is not noetherian [10, p. 179].

Invertible ideals in a noetherian ring are small: since an invertible I is locally principal and regular, $\operatorname{ht} I = 1$ by Krull's PIT. If R is a noetherian integrally closed domain, non-zero principal ideals, and hence also invertible ideals, are even ht 1 unmixed [32, p. 76]; and if R is a UFD, then the invertible ideals are exactly the non-zero principal ideals [32, p. 132].

19. The condition $d(I) < \infty$. Recall that a local ring R, m is called *regular* if $\operatorname{ht} m = \mu(m)$ and that such rings are UFD's and are characterized by the property that for every f.g. R-module N, $d(N) < \infty$ (in fact, $d(N) \leqslant \dim R$). An arbitrary noetherian ring R is called regular if R_m is regular for every maximal ideal m of R.

The condition $d(I) < \infty$ of Theorem 17.1 will usually be satisfied by imposing the requirement that R_p be regular for every $p \in V(I)$. One thing that is true but not immediately evident here is that for a noetherian R one need not additionally assume $\dim R < \infty$ in order to bound the local $d(I_p)$, as the following theorem of Bass-Murthy shows.

19.1. THEOREM [6, p. 26]: *Let R be a noetherian ring and N be a* f.g. *R-module. If $d(N_p) < \infty$ for every prime p, then $d(N) < \infty$.*

Proof: Since $\{p \in \operatorname{Spec} R \mid N_p = 0\} = \operatorname{Spec} R \setminus V(\operatorname{Ann} N)$ when N is finitely generated, this set is open. Claim: N_q is free at a prime q implies N is locally free on a neighborhood of q. Since N_q is free, there exists a f.g. free F and a homomorphism $\phi : F \to N$ such that ϕ_q is an isomorphism. Consider then the exact sequence $0 \to A \to F \overset{\phi}{\to} N \to B \to 0$. Since R is noetherian, A and B are finitely generated; and hence the set $\{p \in \operatorname{Spec} R \mid \phi_p$ is an isomorphism$\} = \{p \in \operatorname{Spec} R \mid A_p = 0\} \cap \{p \in \operatorname{Spec} R \mid B_p = 0\}$ is open by our initial observation. Thus N is free on a neighborhood of q. It now follows that $U_0 = \{p \in \operatorname{Spec} R \mid d(N_p) = 0\}$ is open.

Consider next an exact sequence $0 \to K \to P \to N \to 0$ with P f.g. projective. Again, $U_1 = \{p \in \operatorname{Spec} R \mid d(N_p) \leqslant 1\} = \{p \in \operatorname{Spec} R \mid K_p$ is free$\}$ is open. In this way one constructs a chain of open sets $U_0 \subset U_1 \subset \cdots$. Since $\operatorname{Spec} R$ is a noetherian space, there exists an i such that $U_i = U_{i+1} = \cdots$; and hence $d(N) \leqslant i$. Q.E.D.

20. R-sequences and grade. Let $a_1, \ldots, a_n$ be elements of R such that $(a_1, \ldots, a_n) \neq R$. The sequence $a_1, \ldots, a_n$ is called an *R-sequence* (or a *regular sequence* when it is convenient to omit reference to R) if $a_1 \notin \mathfrak{Z}(R)$, $a_2 \notin \mathfrak{Z}(R/(a_1)), \ldots, a_n \notin \mathfrak{Z}(R/(a_1, \ldots, a_{n-1}))$, or, equivalently, if $a_1 \notin \cup\{p \in \operatorname{Ass}(0)\}$, $a_2 \notin \cup\{p \in \operatorname{Ass}(a_1)\}, \ldots, a_n \notin \cup\{p \in \operatorname{Ass}(a_1, \ldots, a_{n-1})\}$. It follows that if $a_1, \ldots, a_n$ is an R-sequence, then $0 < \operatorname{ht}(a_1) < \operatorname{ht}(a_1, a_2) < \cdots < \operatorname{ht}(a_1, \ldots, a_n)$, and hence $n \leqslant \operatorname{ht}(a_1, \ldots, a_n)$. On the other hand, if $(a_1, \ldots, a_n)$ is a *prime* ideal of ht n in a local ring R, then by 4.1 $a_1, \ldots, a_n$ is an R-sequence.

Any R-sequence contained in an ideal I can be extended to an R-sequence which is maximal with respect to being contained in I; and any two such maximal R-sequences have the same length, denoted $G_R(I)$ and called the *grade* of I. ($G_R(I)$ will be defined to be 0 if I is not regular.) Since associated primes behave well under localization with respect to a m.s. S, i.e., $\operatorname{Ass} I_S = \{pR_S \mid p \in \operatorname{Ass} I$ and $p \cap S = \emptyset\}$, it follows that an R-sequence $a_1, \ldots, a_n$ extends to an R_S-sequence provided $(a_1, \ldots, a_n)R_S \neq R_S$; and therefore $G_R(I) \leqslant G_{R_S}(I_S)$ if $I_S \neq R_S$.

Furthermore, there exists a minimal prime p of I such that $G_R(I) = G_R(p)$. For if $a_1, \ldots, a_n$ is a maximal R-sequence in I, then $I \subset \cup\{\text{primes in } \operatorname{Ass}(a_1, \ldots, a_n)\}$; and since R is noetherian, $\operatorname{Ass}(a_1, \ldots, a_n)$ is finite and therefore $I \subset p \subset q$ for some minimal prime p of I and some $q \in \operatorname{Ass}(a_1, \ldots, a_n)$, and this p works. Note also that $G_R(I) = G_{R_q}(I_q) = n = G_R(q) = G_{R_q}(qR_q)$ since qR_q is an associated prime of $(a_1, \ldots, a_n)R_q$. Thus, it also follows that $G_R(I) = \inf\{G_{R_p}(I_p) \mid p \in V(I)\}$.

From the definition, $G(I) \leqslant \operatorname{ht} I$; and since $\operatorname{ht} I \leqslant \mu(I)$ by Krull's PIT, we always have in a (noetherian) ring

$$G(I) \leqslant \operatorname{ht} I \leqslant \mu(I). \tag{20.1}$$

A number of concepts that will be considered later center on the question of when the equalities hold in 20.1. An important theorem in this respect is the following:

20.2. Theorem (cf. [32, pp. 91 and 93]): *$G(I) = \mu(I)$ (if and) only if I is generated by an R-sequence. Moreover, if I is generated by an R-sequence and $I \subset$ the intersection of the maximal ideals of R, then any minimal generating set for I (i.e., one containing $\mu(I)$ elements) is an R-sequence.*

Another basic theorem, relating grade and f.g. projective dim, is

20.3. Theorem (cf. [32, pp. 125 and 129]): *Let R, m be a local ring and $I \subset m$ be an ideal such that $d(I) < \infty$. Then $G_R(m) = G_{R/I}(m/I) + d_R(R/I)$.*

It may be seen from the Koszul resolution for I (cf. [52]), or by an induction (cf. [32, p. 124]), that if I is generated by an R-sequence of length $n>0$, then $d(I)=n-1$; in fact, the considerations of section III may be viewed as an effort to determine the extent to which an ideal I for which $d(I)=1$ may deviate from being generated by an R-sequence of length 2 locally at primes containing I.

A special case of 20.3 is *Rees's inequality*: $G_R(I) \leqslant d_R(R/I)$. More generally, if $p \in \operatorname{Ass} I$, then $G_R(p) \leqslant d_R(R/I)$. To see this, note that since localization at p at worst increases grade and decreases projective dim, it suffices to consider the case that R is local with maximal ideal p; and in this case 20.3 yields $G_R(p)=d_R(R/I)$. It seems likely that Rees's inequality can be improved to $\operatorname{ht} I \leqslant d_R(R/I)$, which is a special case of the *homological* ht *conjecture* and has been proved for R_{red} containing a field or for I such that $d_R(R/I) \leqslant 2$ (cf. [27], [28]).

In a similar vein, if $d(I)<\infty$, then $G(I)=\inf\{d_{R_p}(R_p/I_p) \mid p \in V(I)\}$. For $G(I) \leqslant G_{R_p}(I_p) \leqslant d_{R_p}(R_p/I_p)$; and on the other hand, there exists $p \in V(I)$ such that $G(I)=G(p)=G_{R_p}(pR_p) \geqslant d_{R_p}(R_p/I_p)$, the inequality following from 20.3.

This last inequality also shows that $G(I)=\operatorname{ht} I$ whenever we have $d_{R_p}(R_p/I_p) \geqslant \operatorname{ht} I_p$, since $\operatorname{ht} I_p \geqslant \operatorname{ht} I$ is always true. Thus, in view of Rees's inequality the homological height conjecture for ideals is equivalent to the following conjecture: If I is an ideal such that $d(I)<\infty$, then $G(I)=\operatorname{ht} I$. In particular, the known case of the homological ht conjecture mentioned above yields the following corollary:

20.4. COROLLARY: *If I is an ideal such that $d(I) \leqslant 1$, then $G(I)=\operatorname{ht} I$.*

21. CM rings and perfect ideals. R will be called a *Cohen-Macaulay ring* (CM for short; Nagata uses "locally Macaulay" and Kaplansky uses merely "Macaulay") if $G(m)=\operatorname{ht} m$ for every maximal ideal m of R; it can be proved that this implies $G(I)=\operatorname{ht} I$ for every ideal $I \neq R$ [32, p. 97]. Classically these rings were studied in terms of the following characterization: R is CM if and only if every c.i. ideal is unmixed (cf. [32, pp. 97 and 115]). For example, any 1-dim noetherian domain or any noetherian, integrally closed domain of dim 2 is CM.

Properties: (a) R is CM implies R_S is CM for every m.s. S of R; and, conversely, if R_m is CM for every maximal ideal m, then R is CM.

(b) R is CM if and only if $R[X]$ is CM [32, p. 109], and every regular R is CM by 4.1. In particular, $k[X_1, \ldots, X_n]$ is CM (where k is a field). Note that there exists a CM (in fact, regular) R having maximal ideals of different ht, e.g., $D[X]$, where D is, say, the ring of integers localized at a prime p.

(c) In a CM ring any two saturated chains of primes descending from a fixed prime have the same length [32, p. 99]. Therefore if R and R/I are CM, a sufficient condition for two minimal primes of I to have the same ht is that they both be contained in some maximal ideal; geometrically, the two irreducible components of $V(I)$ should have a point in common. Moreover, if R, m is CM local and all the minimal primes of I have the same ht, then $\operatorname{ht} m - \operatorname{ht} I = \operatorname{ht}(m/I)$.

As noted previously, for any associated prime p of I, $G_R(I) \leqslant G_R(p) \leqslant d_R(R/I)$. If the equalities hold, the ideal I is called *perfect*. It follows that a perfect ideal is *grade-unmixed*, in the sense that its associated primes all have the same grade. For example, if I is generated by an R-sequence, then I is perfect. Or, if I is a ht 2 ideal of $k[X_1, \ldots, X_n]$, then I is perfect if and only if $d(I) = 1$; the next theorem shows that these I are characterized by the property that they are the unmixed ht 2 ideals for which $k[X_1, \ldots, X_n]/I$ is CM. (For an example of a ht 2 prime p of $k[X_1, \ldots, X_4]$ such that $d(p) = 2$, see [24, p. 179]).

21.1. THEOREM: *Suppose R is* CM *and $I \neq R$ is an ideal such that $d(I) < \infty$. Then the following are equivalent*:

(i) *I is perfect.*

(ii) *I is unmixed and locally perfect at primes containing I.*

(iii) *The minimal primes of I have the same* ht *and R/I is* CM.

Proof: First note that I is perfect implies I is unmixed; for we have already observed that I is grade-unmixed, and grade-unmixed is the same as unmixed in a CM ring. In particular, it suffices to confine our attention to an ideal I such that the minimal primes of I have the same ht.

For any prime $p \supset I$, we then have $G_R(I) = \operatorname{ht} I = \operatorname{ht} I_p = G_{R_p}(I_p) \leqslant d_{R_p}(R_p/I_p) \leqslant d_R(R/I)$. Moreover, since $d(I) < \infty$, there exists a prime $p \supset I$ at which the last inequality becomes equality. The equivalence of (i) and (ii) now follows.

As for (ii)$\Leftrightarrow$(iii), since the CM property is a local one, it suffices to consider the case that R, m is a CM local ring. By 20.3 and the CM property for R, we have $\operatorname{ht} m = G_{R/I}(m/I) + d_R(R/I)$. Then I is perfect $\Leftrightarrow \operatorname{ht} m = G_{R/I}(m/I) + G_R(I) = G_{R/I}(m/I) + \operatorname{ht} I$ $\Leftrightarrow \operatorname{ht} m - \operatorname{ht} I = G_{R/I}(m/I)$. But $\operatorname{ht} m - \operatorname{ht} I = \operatorname{ht}(m/I)$ by 21(c); so I is perfect $\Leftrightarrow \operatorname{ht}(m/I) = G_{R/I}(m/I) \Leftrightarrow R/I$ is CM.

22. Proof of the main theorem (17.1) for R a (noetherian) UFD. It is straightforward to prove theorem 17.1 in the case that R is a UFD, so we shall do this first and then outline in §23 the additional techniques needed to handle the general case.

Before assuming R is a UFD, however, let us first note that we are dealing with an ideal I such that $d(I) > 0$ in 17.1. For in 17.1(a), the hypothesis $\sup\{\mu(I_p) \mid p \in V(I)\} = 2$ implies $d(I) > 0$ because otherwise I would be locally free. In the converse part of 17.1(b), $\operatorname{ht} I^* = 2$ implies $d(I^*) > 0$; and since $I = I_0 I^*$ with I_0 invertible, I is locally isomorphic to I^*, so $d(I^*) > 0$ implies $d(I) > 0$.

Now suppose R is a UFD. Let g be the gcd (greatest common divisor) of the elements of I. Then $I = (g)I^*$, where I^* is a non-zero ideal whose gcd is 1. Moreover, $d(I) > 0$ implies $I^* \neq R$. Choose $a \neq 0$ in I^*, and factor a into irreducible factors: $a = a_1 \cdot \cdots \cdot a_t$. Then gcd $I^* = 1$ implies $I^* \not\subset (a_i)$, $i = 1, \ldots, t$; and hence $I^* \not\subset (a_1) \cup \cdots \cup (a_t)$. Therefore there exists $b \in I^*$ such that a, b have no common irreducible factors, from which it follows that a, b is an R-sequence. Thus, $G(I^*) \geqslant 2$.

Proof of 17.1(a): Suppose $\mu(I) \leqslant 2$ locally. Then $\mu(I^*) \leqslant 2$ locally also; so for any prime $p \supset I^*$ we have $2 \leqslant G(I_p^*) \leqslant \mu(I_p^*) \leqslant 2$. Therefore by 20.2, I_p^* is generated by a regular sequence of length 2, and hence $d(I_p^*) = 1$. Since this is valid locally at every $p \supset I^*$, then $1 = d(I^*) = d(I)$.

Proof of 17.1(b): Suppose $d(I) = 1$. Then $d(I^*) = 1$ also; and therefore $2 \leqslant G(I^*) \leqslant d_R(R/I^*) = 2$. Thus I^* is perfect of grade 2 and hence grade-unmixed, and the special known case of the homological ht conjecture mentioned earlier yields $\operatorname{ht} I^* = 2$. (Note: If R is assumed CM here, then $\operatorname{ht} I^* = 2$ is an immediate consequence of $G(I^*) = 2$.)

Conversely, suppose R is CM and all maximal ideals have ht 3, and suppose $I = I_0 I^*$ with I_0 invertible and I^* ht 2 unmixed. Then $\dim R/I^* = 1$ and no maximal ideal of R/I^* is an associated prime of (0); so R/I^* is CM. By 21.1, I^* is therefore perfect of grade 2, which proves that $d(I^*) = 1$ and hence also that $d(I) = 1$. Q.E.D.

The only use of the UFD property in the above proof was in establishing the decomposition $I = (g)I^*$ with $G(I^*) \geqslant 2$. We shall see in §22 that any ideal I with $0 < d(I) < \infty$ has a decomposition which is locally of this form. We want to note here, however, that if I has a *free* resolution $0 \to R^{s-1} \xrightarrow{\phi} R^s \to I \to 0$, then one can conclude that $I = (g)I^*$ with g a regular element of R and $G(I^*) \geqslant 2$ (provided $d(I) \neq 0$) and, in addition, that I^* is the ideal generated by the $(s-1) \times (s-1)$ minors of the $(s-1) \times s$ matrix (ϕ) (cf. [32, p. 148], [59, p. 861]). Since $d(I^*) = d(I) = 1$, then I^* is perfect, which illustrates a general fact: If H is an $m \times n$ matrix with elements from a noetherian ring R, and if the grade of the ideal I generated by the $(s-1) \times (s-1)$ minors of H is as large as possible, namely $(m-s+2)(n-s+2)$, then I is perfect. The study of such perfect determinantal ideals has a long history involving especially the names of Macaulay, Eagon, and Northcott, and culminating in the recent work of Hochster–J. Roberts [30].

For example, if I is a ht 2 ideal of $R = k[X_1, \ldots, X_n]$ such that $\mu(I) \leqslant 2$ locally, then $d(I) = 1$ by 17.1; and hence by 16.2 and the fact that f.g. R-projectives are free, I has a resolution $0 \to R^{n-1} \xrightarrow{\phi} R^n \to I \to 0$. Since ht $I = 2$, the g of the above remarks must be a unit. Thus, I is the perfect ideal generated by the $(n-1) \times (n-1)$ minors of the matrix (ϕ).

23. MacRae's theorem. We shall show now that, corresponding to the factorization $I = (g)I^*$ of an ideal in a UFD, for a regular ideal $I \neq R$ of an arbitrary (noetherian) R such that $0 < d(I) < \infty$, one has a factorization $I = I_0 I^*$, where I_0 is invertible and $G(I^*) \geqslant 2$. Since the invertible ideal I_0 is locally principal generated by a regular element, then locally $I = (g)I^*$; and therefore the proof of theorem 17.1 given above for a UFD, applied locally, carries over almost verbatim to an arbitrary R. (There remains yet one obstacle to the completion of the proof of 17.1(a) as stated, namely, the above assumption that I is regular. We shall comment later on how to get around this.)

To begin with, invertible ideals must be viewed from a different perspective than that of §18. Let T denote the total quotient ring of R, and for any R-submodule J of T, let $J^{-1} = \{\xi \in T \mid \xi J \subset R\}$. Then $JJ^{-1} \subset R$, and J is called invertible if $JJ^{-1} = R$. It is an easy exercise to verify that an ideal of R is invertible in this sense if and only if it is invertible in the previous sense of being regular and locally principal. If J is invertible, then J^{-1} is also and $(J^{-1})^{-1} = J$. Similarly, since we always have $((J^{-1})^{-1})^{-1} = J^{-1}$, it follows that J^{-1} is invertible if and only if $(J^{-1})^{-1}$ is invertible.

23.1. THEOREM (MacRae [45], Krämer [38]): *If I is a regular ideal such that $d(I) < \infty$, then I^{-1} is invertible.*

The significance of this theorem lies in the fact that if I^{-1} is invertible, then $I = I_0 I^*$, where $I^* = II^{-1}$ and $I_0 = (I^{-1})^{-1}$ are ideals of R such that I_0 is invertible and I^* has the property that $(I^*)^{-1} = R$. The following proposition shows that this last property is equivalent to $G(I^*)$ being $\geqslant 2$ if $I^* \neq R$.

23.2. PROPOSITION: *Let $A \neq R$ be a regular ideal of the noetherian ring R. Then $A^{-1} = R$ if and only if $G(A) \geqslant 2$.*

Proof: $\Rightarrow$: Choose a regular element $a \in A$. If $A/(a) \subset \mathfrak{Z}(R/(a))$, then there exists $b \in R$, $\notin (a)$ such that $bA \subset (a)$; and therefore $b/a \in A^{-1} = R$, which implies $b \in aR$, a contradiction. $\Leftarrow$: If $\xi \in A^{-1}$, we can write $\xi = b/a$, $b \in R$, a regular in A. Since $\xi A \subset R$, $bA \subset aR$. But $G(A) \geqslant 2$ implies there exists $c \in A$ such that c is not a zero-divisor mod (a). Therefore, we must have $b \in aR$, and hence $b/a \in R$. Q.E.D.

If one drops the assumption that I is a regular ideal, it is still possible to factor I so as to carry through the proof of theorem 17.1:

23.3. THEOREM: *Let I be an ideal such that $0 < d(I) < \infty$. Then there exist ideals I_0 and I^* of R such that $I = I_0 I^*$; I_0 is projective (i.e., locally free); $G(I^*) \geqslant 2$; and for any prime p of R, $I_p = 0 \Leftrightarrow (I_0)_p = 0$, and $(I_0)_p = 0$ implies $I_p^* = R_p$.*

Sketch of proof: Decompose R into a finite product of rings having no non-trivial idempotents, $R = R_1 \times \cdots \times R_t$, and let I_j denote the projection of I on R_j. By [32, p. 148], either $I_j = 0$ or I_j is a regular ideal of R_j. If $I_j = 0$, define $(I^*)_j$ to be R_j and $(I_0)_j$ to be 0; if $I_j = R_j$, define $(I^*)_j$ and $(I_0)_j$ to be R_j; and if I_j is regular and $\neq R_j$, choose $(I^*)_j$ and $(I_0)_j$ as in the above discussion. Then let $I^* = (I^*)_1 \times \cdots \times (I^*)_t$ and $I_0 = (I_0)_1 \times \cdots \times (I_0)_t$. Q.E.D.

REMARKS: (a) What we have presented in §23 is actually the cyclic case of MacRae's construction. See also [38], [31], and [59] for some interesting generalizations. While the underlying formulation of theorem 23.1 is due to MacRae, it was Krämer who first emphasized the use of inverses. The idea for removing the regularity assumption on I in 23.3 also comes from Krämer [38, p. 482] (although Krämer's proof ignores the possibility of I localizing to 0, which may happen: let $R = k \times R_1$ and

$I=0\times m_1$, where k is a field and R_1, m_1 is the local ring $k[X,Y]_{(X,Y)}$.

MacRae's theorem also has an interpretation in terms of gcd ideals. Let us call a principal ideal (g) generated by a regular element a *principal* gcd *ideal* for the regular ideal I if (i) $I \subset gR$, and (ii) $I \subset (a/b)R$ for a, b regular in R implies $g \in (a/b)R$. Then define an ideal I_0 of R to be a gcd *ideal* for I if I_0 is locally a principal gcd ideal for I. If such an I_0 exists, it follows that I_0 is locally principal and hence invertible, and that $I_0=(I^{-1})^{-1}$; and, conversely, if I^{-1} is invertible, then the gcd ideal for I exists and equals $(I^{-1})^{-1}$. Thus, an equivalent formulation of MacRae's theorem is that $d(I)<\infty$ implies I has a gcd ideal.

MacRae actually defines an ideal I_0 to be a gcd ideal for I if I_0 locally satisfies (i) and the following weak principal gcd condition, rather than (ii):

(ii′) $I \subset aR$ for a regular in R implies $g \in aR$. However, $d(I)<\infty$ implies $d(rI)<\infty$ for every regular $r \in R$; and (ii′) being valid for all rI, r regular in R, is equivalent to (ii). As a matter of fact, the equivalence of (ii) and (ii′) is the essential step in proving via MacRae's theorem that a regular local ring is a UFD; for, if R is a noetherian domain, then it is easily seen that R is not a UFD if and only if there exists an ideal I generated by two elements having gcd ideal (1) in the sense of (i) and (ii′) but not having a gcd ideal in the sense of (i) and (ii).

(b) Part (a) of theorem 17.1 does not seem to have any generalization to ideals I with $\{\mu(I_p)\}$ bounded by some integer >2, for it is known (cf. [13]) that if there exists a f.g. R-module having projective dim n, then there exists an ideal I of R with $\mu(I)\leqslant 3$ and $d(I)=n-1$.

(c) Sometimes $d(I)=1$ may be seen directly without first verifying that $d(I)<\infty$ as in theorem 17.1. For example, if I is a prime ideal of ht 2 such that $\mu(I)\leqslant 2$ locally, then at any prime $p \supset I$, any minimal generating set for I_p is necessarily a regular sequence by 4.1; and therefore $d_{R_p}(I_p)=1$, which implies $d(I)=1$.

Or, suppose R is an integrally closed domain and I is an ideal of ht 2 such that $\mu(I)\leqslant 2$ locally. Then for any prime $p \supset I$, $2 \leqslant \operatorname{ht} I_p \leqslant \mu(I_p) \leqslant 2$, so it suffices to consider the case that R is local and $\operatorname{ht} I = \mu(I) = 2$. If $I = (a, b)$, then a is regular since R is a domain; and $b \notin \cup\{p \in \operatorname{Ass}(a)\}$ since non-zero principal ideals of a (noetherian) integrally closed domain are ht 1 unmixed [32, p. 76]. Thus, a, b is a regular sequence and $d(I) = 1$.

(d) If $I \neq R$ is a regular ideal of R, then $(I^{-1})^{-1} = \operatorname{Ann} \operatorname{Ext}_R^1(R/I, R)$ (cf. [72, p. 60 and p. 80]). For, applying $\operatorname{Hom}_R(\_, R)$ to the exact sequence $0 \to I \to R \to R/I \to 0$ yields the following diagram with exact rows:

$$\begin{array}{ccccccc}
0 & \to & \operatorname{Hom}_R(R/I, R) & \to & \operatorname{Hom}_R(R, R) & \to \\
 & & \| & & \downarrow & \\
0 & \to & 0 & \to & R & \to
\end{array}$$

$$\begin{array}{ccccc}
\operatorname{Hom}_R(I, R) & \to & \operatorname{Ext}_R^1(R/I, R) & \to & \operatorname{Ext}_R^1(R, R) \\
\downarrow & & \| & & \| \\
I^{-1} & \to & \operatorname{Ext}_R^1(R/I, R) & \to & 0
\end{array}$$

Here the vertical arrows are isomorphisms, the isomorphism $\operatorname{Hom}_R(I, R) \to I^{-1}$ being a consequence of the fact that any element of $\operatorname{Hom}_R(I, R)$ is defined by multiplying I by a suitable element of I^{-1}: given $h \in \operatorname{Hom}_R(I, R)$, if a is a fixed regular element of I and $\xi = h(a)/a$, then for any $x \in I$, $h(a)x = h(x)a$ implies $\xi x = h(x)$. Thus, $I^{-1}/R \cong \operatorname{Ext}_R^1(R/I, R)$; so $(I^{-1})^{-1} = \operatorname{Ann} I^{-1}/R = \operatorname{Ann} \operatorname{Ext}_R^1(R/I, R)$. Interpreted in this light the proposition of §23 asserts that $G(I) \geqslant 2$ if and only if $\operatorname{Ext}_R^1(R/I, R) = 0$, which is a special case of the well-known homological characterization of $G(I)$ as the least n such that $\operatorname{Ext}_R^n(R/I, R) \neq 0$ (cf. [32, p. 101]).

24. Complete intersection ideals and rings. We have previously defined an ideal I of a noetherian ring R to be a c.i. ideal if $\operatorname{ht} I = \mu(I)$. (Classically these ideals have also been called "ideals of the principal class".) This is a slightly weaker notion than that of being generated by an R-sequence, since I is generated by an R-sequence if and only if $G(I) = \operatorname{ht} I = \mu(I)$ (cf. 20.2). An example of a c.i. ideal which is not generated by an R-sequence may be obtained by choosing R, m to be a local ring of $\dim > 0$

such that m is an associated prime of (0) and setting $I=(a)$, $a \in m \setminus \cup \{p \in \operatorname{Spec} R \mid \operatorname{ht} p = 0\}$.

For a CM ring R, the c.i. ideals coincide with the ideals generated by R-sequences. In fact, this property characterizes CM rings: R is CM $\Leftrightarrow$ every c.i. ideal of R is generated by an R-sequence $\Leftrightarrow$ every c.i. ideal of R is unmixed. The second equivalence has been discussed in §21. For the first equivalence, the implication $\Rightarrow$ follows from the fact that $G(I) = \operatorname{ht} I$ in a CM ring, and the reverse implication may be seen as follows: Let m be a maximal ideal of ht d. One can select $a_1, \ldots, a_d \in m$ such that $\operatorname{ht}(a_1, \ldots, a_d) = d$; and since we are given that the c.i. ideal $(a_1, \ldots, a_d)$ is generated by an R-sequence, then the grade of this ideal, and hence also that of m, must be d.

Another hypothesis under which a c.i. ideal I is generated by an R-sequence is $d(I) \leqslant 1$, which follows from 20.4. To emphasize the $d(I) = 1$ case of this observation, we record it as a corollary.

24.1. Corollary: *I is a* c.i. *ideal such that $d(I) = 1$ if and only if I is generated by an R-sequence of length* 2.

A minimal generating set for an ideal generated by an R-sequence need not be itself an R-sequence if R is not local, but Davis (cf. [15]) has put his finger on a property which avoids this anomaly and which at the same time distinguishes nicely between an ideal being a c.i. and being generated by an R-sequence. He calls a sequence of elements $a_1, \ldots, a_n$ of R analytically independent (respectively, strongly analytically independent) if every homogeneous $f(X_1, \ldots, X_n) \in R[X]$ such that $f(a_1, \ldots, a_n) = 0$ necessarily has some power (respectively, the first power) of each of its coefficients in the ideal $(a_1, \ldots, a_n)$; and he proves that an ideal I is a c.i. ideal (respectively, an ideal generated by an R-sequence) $\Leftrightarrow$ some minimal generating set for I is analytically independent (respectively, strongly analytically independent) $\Leftrightarrow$ every minimal generating set for I is analytically independent (respectively, strongly analytically independent). (Grothendieck–Dieudonné use the term "quasi-regular" instead of strongly analytically independent.) Note that analytic independence and strong analytic independence coincide if $a_1, \ldots, a_n$ generate a prime ideal, so it follows that any c.i. prime ideal is generated by an R-sequence (which also follows from 4.1 in case R is local).

A notion related to c.i. ideal is that of *c.i. ring*: a local ring R is called a c.i. ring if there exists a regular local ring $\mathcal{O}$ and a c.i. ideal I of $\mathcal{O}$ such that $R^\wedge \cong \mathcal{O}/I$. Here the completion $R^\wedge$ of R is used instead of R because the completion is always the homomorphic image of some regular local ring, whereas R itself may not be (see [22] for an example of a 1-dim local domain which is not a homomorphic image of a regular local ring). Kiehl–Kunz [35] develop this concept by introducing for the local ring R a non-negative integer $D(R)$ called the *deviation* of R. They prove that $D(R) = D(R^\wedge)$ and that whenever $R \cong \mathcal{O}/I$ with $\mathcal{O}$ regular local, then $D(R) = \mu(I) - \operatorname{ht} I$. It follows that $D(R) = \mu(I) - \operatorname{ht} I$ if $R^\wedge = \mathcal{O}/I$ with $\mathcal{O}$ regular local; and hence R is a c.i. ring if and only if $D(R) = 0$. To illustrate the difficulties involved in passing to the completion, we mention that the following rather innocuous looking question of [35] remains open: If R is a c.i. ring and p is a prime ideal of R, is R_p a c.i. ring? (Avramov [Homology of local flat extensions and complete intersection defects, *Math. Ann.*, **228** (1977), 27–37] has now proved the even stronger result that $D(R_p) \leqslant D(R)$.)

25. The Macaulay examples. These examples show that for any natural number n there exists a ht 2 prime p in $k[X, Y, Z]$, k an algebraically closed field, such that $\mu(p) > n$; in fact, $\mu(p_\theta) > n$ where $\theta = (X, Y, Z)$. The details have been worked out in different ways by Abhyankar [2] and Geyer [23]. The Macaulay examples do not themselves extend to corresponding examples in $k[[X, Y, Z]]$, but Moh [47] has proved the existence of such primes in $k[[X, Y, Z]]$, k any field of characteristic 0, and his primes are extensions of primes from $k[X, Y, Z]$ and thus also prove the existence of such primes there.

J. Sally [60] has raised an interesting question: If $\{\mu(p) \mid p \in \operatorname{Spec} R\}$ is bounded, does it follow that $\dim R \leqslant 2$? If so, then the existence of all of the above examples would be a consequence. As for the converse question, Sally–Vasconcelos [61] point out that there exist noetherian rings R of dim $\leqslant 2$ such that $\{\mu(p) \mid p \in \operatorname{Spec} R\}$ is unbounded, but they offer the following proof that the affine rings are not of this type.

THEOREM: *If R is a finitely generated ring extension of a field k such that* $\dim R \leqslant 2$, *then* $\{\mu(p) \mid p \in \operatorname{Spec} R\}$ *is bounded.*

Proof: Let $R = k[\eta_1, \ldots, \eta_n]$. If p is maximal, then $R/p = k[\xi_1, \ldots, \xi_n]$ is a field and hence the ξ_i are algebraic over k [75, p. 26]. One then verifies that if $f_i(X_1, \ldots, X_i)$, $i = 1, \ldots, n$, is the minimal polynomial for ξ_i over $k[\xi_1, \ldots, \xi_{i-1}]$, then $(f_1(\eta), \ldots, f_n(\eta))$ is maximal and hence equals p.

We may therefore assume p is not maximal. Moreover, since R has only finitely many ht 0 primes, it also suffices to consider the case that R is a domain. By the Noether normalization theorem [75, p. 200], there exist algebraically independent elements $x, y \in R$ such that R is a finite $k[x, y]$ $(= R_0)$-module. Since p is non-maximal and R is integral over R_0, $p \cap R_0$ is also non-maximal [32, p. 29]; or, stated differently, every maximal ideal of R_0 contains an element which does not go to 0 under the composite ring homomorphism $R_0 \to R_0/(p \cap R_0) \to R/p$. Therefore, locally at any maximal ideal of R_0 the grade of R/p as an R_0-module is > 0; and hence by the general version of 20.3 given in [32, p. 125], $d_{R_0}(R/p)$ is locally $\leqslant 1$ and therefore $\leqslant 1$. Now write R as the homomorphic image of a free R_0-module, $\phi : R_0^s \to R$. Then the induced surjective homomorphism $R_0^s \to R/p$ has kernel $\phi^{-1}(p)$, and it follows that $\phi^{-1}(p)$ is a f.g. projective, and hence free, R_0-module. But then $\mu_{R_0}(\phi^{-1}(p)) \leqslant s$, and therefore also $\mu_{R_0}(p) \leqslant s$ and a fortiori $\mu_R(p) \leqslant s$. Q.E.D.

IV. AN ISOMORPHISM RELATING $\operatorname{Ext}^1_R(I, R)$ AND THE MODULE OF DIFFERENTIALS

We have seen in section II how a bound for $\mu(\operatorname{Ext}^1_R(I, R))$ leads to one for $\mu(I)$; for example, by 13.1 if I is a ht 2 unmixed ideal of $R = k[X, Y, Z]$, then $\mu(I) = 2$ if and only if $\operatorname{Ext}^1_R(I, R)$ is a cyclic

R-module. We want to show next that for the ideal I of a non-singular unmixed variety of $\dim n-2$ in $\mathscr{C}^n$, $\operatorname{Ext}_R^1(I,R)$ depends only on the affine ring R/I. This will follow from the isomorphisms $\operatorname{Ext}_R^1(I,R) \cong \operatorname{Hom}_{R/I}(\bigwedge^2(I/I^2), R/I) \cong \bigwedge^{n-2}\Omega_k(R/I)$, where $\Omega_k(R/I)$ denotes the module of k-differentials of R/I.

Fix a ring R and an ideal $I \neq R$. In §§26–28, I will be assumed to be generated by an R-sequence of length 2, i.e., $I=(a_1,a_2)$, where a_1 is a regular element of R and a_2 is a regular element $\operatorname{mod}(a_1)$; and it will be shown how such an R-sequence yields an isomorphism between $\operatorname{Ext}_R^1(I,R)$ and $\operatorname{Hom}_{R/I}(\bigwedge^2(I/I^2), R/I)$. Then in §29 we shall begin with an ideal I which is locally generated by a regular sequence of length 2 at every prime $p \supset I$ and shall first show that for a noetherian ring R this property can be spread out to a neighborhood of p. These neighborhood regular sequences will be used to define neighborhood isomorphisms which agree on overlapping neighborhoods, and then the neighborhood isomorphisms will be patched together to get the required global isomorphism; of course, in carrying out this program it is essential to keep track of how each isomorphism is defined. Finally, in §30 the further isomorphism

$$\operatorname{Hom}_{R/I}(\textstyle\bigwedge^2(I/I^2), R/I) \cong \Omega_k(R/I)$$

will be discussed and some immediate applications given.

26. The Koszul resolution. (Cf. [52, p. 351], [32, p. 128].) Suppose I is generated by a regular sequence a_1, a_2. The Koszul resolution with respect to a_1, a_2, denoted $\mathcal{K}(a_1,a_2)$, is the sequence $(\phi,\psi): 0 \to R \to R^2 \to I \to 0$, where ψ is defined by $\psi(e_i)=a_i$ and ϕ is defined by $\phi(e)=(a_2,-a_1) \in R^2$. (Here $e_1=(1,0)$ and $e_2=(0,1)$ in R^2, and $e=(1)$ in R.) This sequence is exact, which amounts to saying that $(a_2,-a_1)$ is a free generator for $\ker\psi$. The freeness is a consequence of a_1 being regular. To see that $(a_2,-a_1)$ generates $\ker\psi$, let $(b_1,b_2) \in \ker\psi$ and note that $b_1a_1+b_2a_2=0$ implies $b_2a_2 \equiv 0 \operatorname{mod}(a_1)$. But then $b_2 \equiv 0 \operatorname{mod}(a_1)$ since a_2 is regular $\operatorname{mod}(a_1)$; so $b_2=ca_1$, $c \in R$. Then $(b_1+ca_2)a_1=0$, which implies $b_1+ca_2=0$ since a_1 is regular. Therefore, $(b_1,b_2)=-c(a_2,-a_1)$.

Usually the generator e of R in the resolution $\mathcal{K}(a_1, a_2)$ is denoted $e_1 \wedge e_2$, so that ϕ is defined by $\phi(e_1 \wedge e_2) = (a_2, -a_1)$. This notation comes from exterior algebra and is suggestive of what happens under a change of R-sequence. Thus, suppose a_1', a_2' is another R-sequence which generates I and is related to a_1, a_2 by $a_i = r_{i1}a_1' + r_{i2}a_2'$, $i = 1, 2$, and $r_{ij} \in R$. If $\delta : R^2 \to R^2$ is the homomorphism defined by $\delta(e_i) = r_{i1}e_1' + r_{i2}e_2'$, then there exists a unique homomorphism $\alpha : R \to R$ which makes the diagram

$$\begin{array}{lccccccc} \mathcal{K}(a_1', a_2'): & 0 \to & R & \to & R^2 \to & I \to 0 \\ & & \alpha \uparrow & & \delta \uparrow & \| \\ \mathcal{K}(a_1, a_2): & 0 \to & R & \to & R^2 \to & I \to 0 \end{array}$$

commutative, and α is defined by $\alpha(e) = re'$ for some $r \in R$. Let us compute this r: $r(a_2', -a_1') = \delta((a_2, -a_1)) = a_2(r_{11}e_1' + r_{12}e_2') - a_1(r_{21}e_1' + r_{22}e_2') = (a_2 r_{11} - a_1 r_{21}, a_2 r_{12} - a_1 r_{22})$. Therefore

$$\begin{aligned} -ra_1' &= a_2 r_{12} - a_1 r_{22} = (r_{21}a_1' + r_{22}a_2')r_{12} - (r_{11}a_1' + r_{12}a_2')r_{22} \\ &= (r_{21}r_{12} - r_{11}r_{22})a_1'. \end{aligned}$$

Thus, $r = r_{11}r_{22} - r_{21}r_{12} = \det(r_{ij})$ since a_1' is regular. One consequence is that, in view of how scalar multiplication is defined for $\mathrm{Ext}_R^1(I, R)$ (cf. §11), we have $\det(r_{ij}) \cdot [\mathcal{K}(a_1, a_2)] = [\mathcal{K}(a_1', a_2')]$, where [ ] denotes equivalence class in $\mathrm{Ext}_R^1(I, R)$.

27. $\mathrm{Ext}_R^1(I, R) \cong R/I$. We now want to define an isomorphism $\rho(a_1, a_2) : \mathrm{Ext}_R^1(I, R) \to R/I$, where ρ depends on the choice of R-sequence a_1, a_2 generating I. Applying $\mathrm{Hom}_R(\_, R)$ to the Koszul resolution $\mathcal{K}(a_1, a_2) : 0 \to R \xrightarrow{\phi} R^2 \xrightarrow{\psi} I \to 0$ gives rise to the following commutative diagram:

$$\begin{array}{ccccccc} \mathrm{Hom}_R(R^2, R) & \xrightarrow{\nu} & \mathrm{Hom}_R(R, R) & \xrightarrow{\eta} & \mathrm{Ext}_R^1(I, R) & \to 0 \\ \| & & \tau \downarrow & & & \\ \mathrm{Hom}_R(R^2, R) & \to & R & \to & R/I & \to 0 \end{array}$$

The top row is exact by §11 (or one may also apply [44, p. 89]), the map τ is by definition the R-module isomorphism defined by evaluating any $h \in \operatorname{Hom}_R(R, R)$ at the generator e of R, and the map $R \to R/I$ is the canonical map. Recall also from §11 that $\eta(\mathrm{id}) = [\mathcal{K}(a_1, a_2)]$ and thus that $[\mathcal{K}(a_1, a_2)]$ generates $\operatorname{Ext}_R^1(I, R)$.

Let us verify now that the bottom row of the diagram is also exact, which amounts to checking that $I = \operatorname{Im}(\tau \circ \nu)$. If $r \in \operatorname{Im}(\tau \circ \nu)$, then there exists $h \in \operatorname{Im} \nu$ such that $r = h(e)$. Note that $h \in \operatorname{Im} \nu$ if and only if there exists $h' \in \operatorname{Hom}_R(R^2, R)$ such that $h' \circ \phi = h$, by the definition of ν. Therefore $r = h(e) = h' \circ \phi(e) = h'(a_2 e_1 - a_1 e_2) = a_2 h'(e_1) - a_1 h'(e_2) \in I$. Conversely, if $r \in I$, then $r = b_1 a_2 - b_2 a_1, b_i \in R$; and we can define $h' \in \operatorname{Hom}(R^2, R)$ by $h'(e_i) = b_i$. Then $h = h' \circ \phi \in \operatorname{Im} \nu$ and $h(e) = h'(a_2 e_1 - a_1 e_2) = a_2 b_1 - a_1 b_2 = r$; so $r \in \operatorname{Im}(\tau \circ \nu)$.

It now follows that the above diagram may be augmented by a homomorphism $\rho : \operatorname{Ext}_R^1(I, R) \to R/I$ which is uniquely determined by the requirement that the resulting diagram remain commutative. Moreover, ρ is explicitly described by $\rho([\mathcal{K}(a_1, a_2)]) = \mathrm{id}$ of R/I; and since τ is an isomorphism, ρ must also be an isomorphism.

What happens if one uses a different R-sequence a_1', a_2'? Then one obtains an isomorphism $\rho' : \operatorname{Ext}_R^1(I, R) \to R/I$ defined by $\rho'([\mathcal{K}(a_1', a_2')]) = \mathrm{id}$ of R/I. Since we have seen in §26 that $[\mathcal{K}(a_1', a_2')] = \det(r_{ij})[\mathcal{K}(a_1, a_2)]$, where $a_i = r_{i1} a_1' + r_{i2} a_2'$, $i = 1, 2$, then $\rho = \det(r_{ij})\rho'$.

28. $R/I \cong \operatorname{Hom}_{R/I}(\wedge^2(I/I^2), R/I)$. We continue to assume that I is generated by an R-sequence a_1, a_2.

28.1. LEMMA: *If an ideal J of a ring R is generated by an R-sequence $a_1, \ldots, a_t$, then J/J^2 is (R/J)-free on $a_1^*, \ldots, a_t^*$ (where a_i^* denotes the canonical image of a_i in J/J^2).*

Proof: We must show that $r_1 a_1 + \cdots + r_t a_t \in J^2$, $r_i \in R$, implies $r_1, \ldots, r_t \in J$. If $t = 1$, this is immediate since $a_1 \notin \mathcal{Z}(R)$; so assume $t > 1$ and proceed by induction on t. Since $r_1 a_1 + \cdots + r_t a_t \in J^2$, there exist $r_i' \in R$ such that $r_1' a_1 + \cdots + r_t' a_t = 0$ and $r_i' \equiv r_i \bmod J$. Since $a_t \notin \mathcal{Z}(R/(a_1, \ldots, a_{t-1}))$, then $r_t' \in (a_1, \ldots, a_{t-1}) \subset J$; and hence there exist $r_i'' \in R$ such that

$r_1''a_1 + \cdots + r_{t-1}''a_{t-1} = 0$ and $r_i'' \equiv r_i' \equiv r_i \mod J$. Now by the induction hypothesis applied to the ideal $(a_1, \ldots, a_{t-1})$, $r_1'', \ldots, r_{t-1}'' \in (a_1, \ldots, a_{t-1}) \subset J$. Q.E.D.

We shall see in 35.2 that the converse to 28.1 is valid for an ideal J of a local ring R such that $d(J) < \infty$.

A consequence of the above lemma is that $\bigwedge^2(I/I^2)$ is R/I-free on $a_1^* \wedge a_2^*$ (cf. [9, p. 66]), and hence the evaluation map $\sigma(a_1, a_2) : \mathrm{Hom}_{R/I}(\bigwedge^2(I/I^2), R/I) \to R/I$ defined by mapping $h \in \mathrm{Hom}_{R/I}(\ ,\ )$ to $h(a_1^* \wedge a_2^*)$ is an R/I-module isomorphism. Again we want to know what happens to $\sigma = \sigma(a_1, a_2)$ when one begins with another regular sequence a_1', a_2', related to a_1, a_2 by $a_i = r_{i1}a_1' + r_{i2}a_2'$. Since $a_1^* \wedge a_2^* = \det(r_{ij}^*)(a_1'^* \wedge a_2'^*)$ (cf. [9, p. 78]), it follows that $\sigma(h) = h(a_1^* \wedge a_2^*) = \det(r_{ij}^*)h(a_1'^* \wedge a_2'^*) = \det(r_{ij}^*)$ $\sigma'(h)$, where $\sigma' = \sigma(a_1', a_2')$. Thus, $\sigma = \det(r_{ij}^*)\sigma'$.

One can now combine the isomorphisms of §§27 and 28 to get an isomorphism

$$\sigma^{-1} \circ \rho : \mathrm{Ext}_R^1(I, R) \xrightarrow{\rho} R/I \xrightarrow{\sigma^{-1}} \mathrm{Hom}_{R/I}\big(\textstyle\bigwedge^2(I/I^2), R/I\big).$$

(It does not matter here whether one considers these as R-modules or R/I-modules.) This isomorphism $\sigma^{-1} \circ \rho$ is defined with respect to an R-sequence a_1, a_2 of length 2 which generates I. However, we have also seen in §§27 and 28 that if a_1', a_2' is another R-sequence which generates I and is related to a_1, a_2 by $a_i = r_{i1}a_1' + r_{i2}a_2'$, then $\sigma^{-1} \circ \rho = [1/\det(r_{ij}^*)](\sigma')^{-1} \circ [\det(r_{ij})]\rho'$ $= (\sigma')^{-1} \circ \rho'$; so the composite isomorphism is actually independent of the R-sequence chosen.

29. The global isomorphism.

The following lemma shows that the property of being generated by a regular sequence locally at a prime p can be spread out to a neighborhood of p.

LEMMA: *Let R be a noetherian ring and $I \neq R$ be an ideal of R, and let $p \supset I$ be a prime ideal of R. If I_p is generated by a regular sequence of length n, then there exists $s \notin p$ such that I_s is generated by a regular sequence of length n.*

Proof: By hypothesis there exist $a_1, \ldots, a_t \in I$ such that $I_p = (a_1/1, \ldots, a_t/1)$ and $a_1/1, \ldots, a_t/1$ is an R_p-sequence. Since I is f.g., there exists $s_0 \notin p$ such that $s_0 I \subset (a_1, \ldots, a_t)$. Moreover, $\{q \in \mathrm{Ass}(a_1, \ldots, a_{i-1}) \mid q \not\subset p\}$ is finite; so there exists $s_i \notin p$ such that $s_i \in \cap \{q \in \mathrm{Ass}(a_1, \ldots, a_{i-1}) \mid q \not\subset p\}$, $i = 1, \ldots, t$. Now let $s = s_0 s_1 \cdot \cdots \cdot s_t$, and then the canonical images of $a_1, \ldots, a_t$ in R_s are the required generators. Q.E.D.

The next proposition is the tool that enables one to patch together a collection of neighborhood homomorphisms to a global homomorphism.

29.1. PROPOSITION: *Let $s_1, \ldots, s_n$ be elements of a ring R such that $(s_1, \ldots, s_n) = R$; let M, N be R-modules; and suppose there are given R_{s_i}-module homomorphisms $\phi_i : M_{s_i} \to N_{s_i}$, $i = 1, \ldots, n$. For $i \neq j$, let ϕ_{ij} be the $R_{s_i s_j}$-module homomorphism which makes the following diagram commutative:*

$$\begin{array}{ccc} M_{s_i s_j} & \xrightarrow{\phi_{ij}} & N_{s_i s_j} \\ \uparrow & & \uparrow \\ M_{s_i} & \xrightarrow{\phi_i} & N_{s_i} \end{array} . \tag{29.2}$$

(Here the vertical maps are the canonical homomorphisms.) If $\phi_{ij} = \phi_{ji}$ for $i \neq j$, then there exists a homomorphism $\phi : M \to N$ such that $\phi_{s_i} = \phi_i$ for $i = 1, \ldots, n$.

The proposition is undoubtedly well known, especially to sheaf-theorists, but I know of no well-known reference for it as stated. The essential ingredient for defining the map ϕ is the following (cf. [5, p. 3.3]): Let N be an R-module, and suppose there are given elements $s_1, \ldots, s_n \in R$ such that $(s_1, \ldots, s_n) = R$ and elements $a_i \in N_{s_i}$, $i = 1, \ldots, n$, such that the canonical images of a_i and a_j in $N_{s_i s_j}$ coincide for all $i \neq j$. Then there exists a unique element $a \in N$ such that the canonical image of a in N_{s_i} is a_i for $i = 1, \ldots, n$.

Before applying the proposition, we shall make a few comments on its content:

(a) Geometrically, each s_i defines an open subset U_i of $\operatorname{Spec} R$, namely, $U_i = \{p \in \operatorname{Spec} R \mid s_i \notin p\}$; and the assertion $(s_1, \ldots, s_n) = R$ is equivalent to saying that $U_1 \cup \cdots \cup U_n = \operatorname{Spec} R$. The map $\phi_i : M_{s_i} \to N_{s_i}$ should be thought of as being defined on U_i and the equality $\phi_{ij} = \phi_{ji}$ as saying that ϕ_i and ϕ_j agree on the overlap $U_i \cap U_j$. The proposition then asserts that there exists a global homomorphism ϕ defined on all of $\operatorname{Spec} R$ and which agrees with ϕ_i on U_i.

(b) If each ϕ_i is an isomorphism, then ϕ is also; for then $\operatorname{coker} \phi$ and $\ker \phi$ are locally 0 at every prime p and hence are 0.

(c) If I is an ideal of R such that $I \subset \operatorname{Ann} M \cap \operatorname{Ann} N$, so that both M and N may be regarded as R/I-modules, then by applying the proposition with respect to R/I it suffices to assume $(s_1, \ldots, s_n, I) = R$ rather than $(s_1, \ldots, s_n) = R$.

(d) The module $M_{s_i s_j}$ is canonically isomorphic to $(M_{s_i})_{s_j}$ by mapping $m/s_i s_j$ to $(m/s_i)/s_j$. The significance of this is that $M_{s_i s_j}$ may be regarded as the localization of M_{s_i} at s_j, even though, strictly speaking, it is not.

(e) Since $\phi_{ij} = 0 = \phi_{ji}$ if $s_i s_j$ is nilpotent, one need only be concerned with those ϕ_{ij} such that $s_i s_j$ is not nilpotent. Geometrically this means that one need not worry about neighborhoods that have no overlap. Moreover, if one passes to R/I as in (c), one need only be concerned with those ϕ_{ij} such that $s_i s_j \notin \sqrt{I}$, or, equivalently, such that $I_{s_i s_j} \neq R_{s_i s_j}$.

29.3. THEOREM: *Let R be a noetherian ring and $I \neq R$ be an ideal of R such that I is locally generated by a regular sequence of length 2 at every prime containing I. Then* $\operatorname{Ext}^1_R(I, R) \cong \operatorname{Hom}_{R/I}(\bigwedge^2(I/I^2), R/I)$.

Proof: For each prime $p \supset I$, there exists $s = s(p)$ in $R \setminus p$ such that I_s is generated by a regular sequence, by the above lemma. Then no prime contains I and all of these $s(p)$, so $(\{s(p)\}, I) = R$. Therefore there exist finitely many $s(p)$, say $s_1, \ldots, s_n$, such that $(s_1, \ldots, s_n, I) = R$.

Let a_i, b_i be the regular R_{s_i}-sequence which generates I_{s_i}; and let a_{ij}, b_{ij} denote the canonical images of a_i, b_i in $R_{s_i s_j}$. By §28, there exist isomorphisms Γ_i and Γ_{ij} (whenever $I_{s_is_j} \neq R_{s_is_j}$) defined with respect to a_i, b_i and a_{ij}, b_{ij}, respectively, which may be seen to make the following diagram commutative. (The notation becomes a bit unwieldy at this point, so let us simplify by dropping the s and merely writing R_i, I_i, R_{ij}, etc., instead of $R_{s_i}, I_{s_i}, R_{s_is_j}$.)

$$\begin{array}{ccc} \mathrm{Ext}^1_{R_{ij}}(I_{ij}, R_{ij}) & \overset{\Gamma_{ij}}{\rightarrow} & \mathrm{Hom}_{R_{ij}/I_{ij}}\left(\bigwedge^2(I_{ij}/I_{ij}^2), R_{ij}/I_{ij}\right) \\ \uparrow & & \uparrow \\ \mathrm{Ext}^1_{R_i}(I_i, R_i) & \overset{\Gamma_i}{\rightarrow} & \mathrm{Hom}_{R_i/I_i}\left(\bigwedge^2(I_i/I_i^2), R_i/I_i\right) \end{array} \tag{29.4}$$

Moreover, $\Gamma_{ij} = \Gamma_{ji}$ since these isomorphisms were seen in §28 to be independent of the regular sequences used to define them. Thus, the theorem follows from the proposition once one verifies that after making some canonical identifications this diagram becomes the localization diagram of the proposition. Q.E.D.

The material of §§26–30 may be found in [4], in greater generality but with less detail. One can prove the following generalization of theorem 29.3 by a straightforward generalization of the given argument:

29.4. THEOREM: *If t is an integer $\geqslant 2$ and $I \neq R$ is an ideal of the noetherian ring R such that I is locally generated by a regular sequence of length t at every prime containing I, then* $\mathrm{Ext}^{t-1}_R(I, R) \cong \mathrm{Hom}_{R/I}(\bigwedge^t(I/I^2), R/I)$.

30. $\mathrm{Hom}_{R/I}(\bigwedge^2(I/I^2), R/I) \cong \Omega_k(R/I)$ for the ideal of a non-singular space curve. We shall first list a few elementary properties of $\otimes$ and $\bigwedge$ that will be needed (cf. [9] and [10]):

(a) If A, B are ideals of a ring R, then there exists a canonical surjective R-linear map $h: A \otimes_R B \rightarrow AB$ taking $a \otimes b$ to ab. Moreover, if B is flat, then h is an isomorphism, as one sees by tensoring the sequence $0 \rightarrow A \rightarrow R$ with B.

(b) If A, B are R-modules, then $\bigwedge^n(A \oplus B) \cong \bigoplus_{i=0}^n[(\bigwedge^{n-i}A) \otimes_R (\bigwedge^i B)]$, where $\bigwedge^0 A$ is defined to be R and $\bigwedge^1 A$ to be A. Moreover, if t is an integer $\geqslant 0$ and A is free of rk t, then

$\bigwedge^t A \cong R$ and $\bigwedge^{t+i} A = 0$ for $i \geqslant 1$. Since $\bigwedge^t_{R_S} A_S \cong (\bigwedge^t A)_S$ for any m.s. S of R, this last remark applied locally shows that if A is locally free of constant rk t, then $\bigwedge^t A$ is locally free of constant rk 1 and $\bigwedge^{t+i} A = 0$ for $i \geqslant 1$.

(c) Suppose A is a f.g. projective R-module of constant rk t. Then there exists a "complementary" R-module B such that $A \oplus B \cong R^n$ for some n, and B is f.g. projective of constant rk $n - t$. If, in addition, R is noetherian, then $\bigwedge^{n-t} B \cong \mathrm{Hom}(\bigwedge^t A, R)$, which may be seen as follows: $(\bigwedge^t A) \otimes (\bigwedge^{n-t} B) \cong R$ and $\bigwedge^t A, \bigwedge^{n-t} B$ are f.g. projectives of constant rk 1 by (b). Therefore $\bigwedge^t A, \bigwedge^{n-t} B$ are isomorphic to invertible ideals, P, Q, respectively, of R by 18.1. Moreover, $P \otimes Q \cong R$ implies $PQ \cong R$ by (a); and hence $PQ = rR$ for some regular $r \in R$. Therefore $Q = rP^{-1} \cong P^{-1}$, and $P^{-1} \cong \mathrm{Hom}(P, R)$. (This last isomorphism is explained in Remark (d) following §24.) Q.E.D.

Combining (c) with 28.1 and 29.4, we have

30.1. THEOREM: *Suppose $I \neq R$ is an ideal of a noetherian ring R such that I is locally generated by a regular sequence of fixed length $t \geqslant 1$ at every prime containing I. Then*

(i) *I/I^2 is R/I-projective of constant* rk t;

(ii) *if $t \geqslant 2$, then $\mathrm{Ext}_R^{t-1}(I, R) \cong \mathrm{Hom}_{R/I}(\bigwedge^t(I/I^2), R/I)$; and*

(iii) *if B is an R/I-module such that $(I/I^2) \oplus B \cong (R/I)^n$, then $\mathrm{Hom}_{R/I}(\bigwedge^t(I/I^2), R/I) \cong \bigwedge^{n-t} B$.*

Note: Assertion (ii) also holds for $t = 1$ if one uses $\mathrm{Ext}_R^t(R/I, R)$ instead of $\mathrm{Ext}_R^{t-1}(I, R)$. (For $t \geqslant 2$, $\mathrm{Ext}_R^{t-1}(I, R) \cong \mathrm{Ext}_R^t(R/I, R)$; but this is not usually valid for $t = 1$.)

The module that will play the role of the complementary module B in this theorem is the module of differentials $\Omega_k(R/I)$, to be introduced in the next section. If $R = k[X_1, \ldots, X_n]$, k a perfect field, and I is a ht t unmixed ideal such that R/I is regular, then we shall see (33.4) that $(I/I^2) \oplus \Omega_k(R/I) \cong (R/I)^n$; and therefore by 30.1, $\mathrm{Ext}_R^{t-1}(I, R) \cong \bigwedge^{n-t} \Omega_k(R/I)$. The only case of this isomorphism that will be useful to us is that of $n = 3$ and $t = 2$:

30.2. Corollary: *Suppose I is a* ht 2 *unmixed ideal of* $R = k[X, Y, Z]$, *k a perfect field, such that R/I is regular. Then* $\mathrm{Ext}^1_R(I, R) \cong \Omega_k(R/I)$.

To carry this line of thought a bit further, we have seen in 13.2 that the ideal I of a non-singular curve in $\mathcal{Q}^3$ is a complete intersection if and only if $\mathrm{Ext}^1_R(I, R)$ is a cyclic R-module, and hence in view of 30.2, if and only if $\Omega_k(R/I)$ is a cyclic R/I-module. We can now extend this to the ideal of a non-singular curve in $\mathcal{Q}^n$:

30.3. Corollary: *Let I be a* ht $n-1$ *unmixed ideal in* $R = k[X_1, \ldots, X_n]$, *k an algebraically closed field, such that R/I is regular. Then the following are equivalent:*

(i) *$\Omega_k(R/I)$ is a* rk 1 *free R/I-module.*
(ii) *$\Omega_k(R/I)$ is a cyclic R/I-module.*
(iii) *I is a complete intersection.*

Proof: (i)$\Rightarrow$(ii): Immediate.
(ii)$\Rightarrow$(iii): By the remarks of §6 we may assume $n = 3$ (this is where the algebraic closure of k is needed). Then $\Omega_k(R/I) \cong \mathrm{Ext}^1_R(I, R)$ by 30.2, and hence I is a c.i. by 13.2.
(iii)$\Rightarrow$(i): I/I^2 is R/I-free of rk $n-1$ by 28.1. Therefore $\bigwedge^{n-1}(I/I^2) \cong R/I$, and consequently $\mathrm{Hom}_{R/I}(\bigwedge^{n-1}(I/I^2), R/I) \cong R/I$. Now apply 30.1 (iii) and the remarks following it.

30.4. Corollary: *If a non-singular curve C in $\mathcal{Q}^n$ is a complete intersection, then so also is every curve isomorphic to C.*

V. THE MODULE OF DIFFERENTIALS

Fix throughout section V rings k and A and a homomorphism $k \to A$ making A a k-algebra. (The case to keep in mind is that of a field $k \subset A$.)

31. Definition and properties of $\Omega_k(A)$. (Cf. [40], [46].) A *k-derivation* of the k-algebra A into an A-module M is a k-linear map $\delta : A \to M$ such that $\delta(ab) = a\delta(b) + b\delta(a)$ for all $a, b \in A$.

Note that $\delta(\xi \cdot 1_A) = 0$ for $\xi \in k$; and, conversely, if δ is a Z-derivation (Z = integers) of A such that $\delta(\xi \cdot 1_A) = 0$ for all $\xi \in k$, then δ is a k-derivation.

The *universal k-derivation* for A is defined to be a pair $(d, \Omega_k(A))$, consisting of an A-module $\Omega_k(A)$ and a k-derivation $d: A \to \Omega_k(A)$ such that for any k-derivation $\delta: A \to M$ there exists a unique A-linear map $h: \Omega_k(A) \to M$ such that $h \circ d = \delta$:

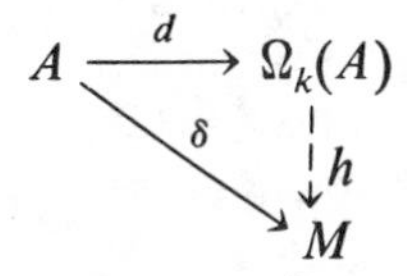

The module $\Omega_k(A)$ is determined up to isomorphism and is called the *module of k-differentials* of A; it is generated as an A-module by $\{da \mid a \in A\}$. The dual of $\Omega_k(A)$, $\mathrm{Hom}_A(\Omega_k(A), A)$, is isomorphic to the module of all k-derivations of A into A.

31.1. EXAMPLE: Let $k \subset K$ be fields with K finitely generated over k. Then K is separably generated over k if and only if $\mu(\Omega_k(K)) = \mathrm{tr.deg.}(K/k)$. Moreover, the differentials of the elements of a separating transcendence basis of K over k form a minimal generating set for $\Omega_k(K)$. In particular, K is separably algebraic over k if and only if $\Omega_k(K) = 0$. The proof of these statements rests on an examination of how a derivation of a field into itself extends to a simple extension field (cf. [73, p. 13] or [46, p. 190]).

Given a k-algebra A' and a k-algebra homomorphism $\phi: A \to A'$, there exists a unique A-linear map $\phi_*: \Omega_k(A) \to \Omega_k(A')$ such that $\phi_* \circ d = d' \circ \phi$. (Here $\Omega_k(A')$ is to be considered an A-module via its A'-module structure and the map ϕ:

$$\begin{array}{ccccc} & \nearrow & A & \longrightarrow & \Omega_k(A) \\ k & & \downarrow \phi & & \downarrow \phi_* \\ & \searrow & A' & \longrightarrow & \Omega_k(A').) \end{array}$$

In particular, if I is an ideal of A and ϕ is the canonical homomorphism $A \to A/I$, then ϕ_* is surjective and has kernel

$\langle dI \rangle$, the submodule of $\Omega_k(A)$ generated by $\{da \mid a \in I\}$; and therefore $\Omega_k(A/I) \cong \Omega_k(A)/\langle dI \rangle$. This isomorphism is an A-module isomorphism by definition; but since $I\Omega_k(A) \subset \langle dI \rangle$, it may also be considered an A/I-module isomorphism. If now the exact sequence $0 \to \langle dI \rangle \to \Omega_k(A) \to \Omega_k(A/I) \to 0$ is tensored with A/I, then the resulting exact sequence may be written in the form

$$I/I^2 \to \Omega_k(A) \otimes_A A/I \to \Omega_k(A/I) \to 0, \tag{31.2}$$

where the A/I-homomorphism $I/I^2 \to \Omega_k(A) \otimes_A A/I$ is the one induced by the derivation map of I into $\Omega_k(A)$.

The module of differentials $\Omega_k(A)$ may, of course, be constructed by forming the free A-module on the symbols $\{da \mid a \in A\}$ and then dividing out by the submodule generated by the minimal collection of relations needed to make the map taking $a \in A$ to the equivalence class of da a k-derivation. However, there is a more useful construction: First write A as the homomorphic image of a polynomial ring $R = k[\{X_\lambda\}]$, say $A = R/I$. One checks that $\Omega_k(R)$ is just the free R-module on the symbols dX_λ and that the universal k-derivation d for R is the map taking $f \in R$ to $\sum_\lambda (\partial f/\partial X_\lambda) dX_\lambda$. Then $\Omega_k(R) \otimes_R R/I \cong \bigoplus_\lambda A dX_\lambda$, the free A-module on the symbols dX_λ; and by the above exact sequence $\Omega_k(A) \cong (\bigoplus_\lambda A dX_\lambda)/K$, where K is the A-submodule of $\bigoplus_\lambda A dX_\lambda$ generated by $\{\sum_\lambda (\partial f/\partial x_\lambda) dX \mid f \in I\}$, $\partial f/\partial x_\lambda$ denoting the canonical image of $\partial f/\partial X_\lambda$ in R/I. Moreover, if a particular generating set $\{f_j\}$ for I is specified, then K is generated by $\{\sum_\lambda (\partial f_j/\partial x_\lambda) dX_\lambda\}$; and hence the matrix associated with this generating set for K is just $\mathcal{J}(f_j; x) = (\partial f_j/\partial x_\lambda)$. One consequence of the discussion of Fitting ideals to be taken up next is that the rank of this matrix depends only on R/I.

REMARK: It follows from the above that if A is a f.g. k-algebra, then $\Omega_k(A)$ is a f.g. A-module. More generally, if A is a localization of a f.g. k-algebra A' with respect to a m.s. S of A', then $\Omega_k(A)$ is a f.g. A-module; for $\Omega_k(A'_S) \cong \Omega_k(A')_S$. The general question of when $\Omega_k(A)$ is finitely generated is difficult, even for a DVR (cf. [8]).

32. Rk Ω. Example 31.1 may be used to compute $\operatorname{rk}\Omega_k(A)$ under certain conditions. For example, suppose k is a field and A is a f.g. k-algebra; and assume in addition that A is a domain whose quotient field $A_{(0)}$ is separably generated over k. Then $\operatorname{rk}\Omega_k(A) = \mu(\Omega_k(A)_{(0)}) = \mu(\Omega_k(A_{(0)})) = \operatorname{tr.deg.}(A_{(0)}/k) = \dim A$, where the first equality is by definition, the second by commutativity of $\Omega_k(\;)$ and localization, the third by 31.1, and the fourth by the main theorem of dimension theory [75, p. 193]. More generally,

32.1. PROPOSITION: *Let k be a field and A be a reduced* f.g. *k-algebra such that A_p is separably generated over k for every* ht 0 *prime p of A.* (Note that A_p is the quotient field of A/p.) *Then* $\operatorname{rk}\Omega_k(A) = \inf\{\dim A/p \mid \operatorname{ht} p = 0\}$.

Proof: Let $\Omega = \Omega_k(A)$. By definition $\operatorname{rk}\Omega = \inf\{\mu(\Omega_p) \mid \operatorname{ht} p = 0\}$. But for any ht 0 prime p of a reduced ring A, $A_p = (A/p)_{(0)}$; so by the commutativity of $\Omega_k(\;)$ and localization, $\Omega_p = \Omega_k(A_p) = \Omega_k(A/p)_{(0)}$. Therefore $\mu(\Omega_p) = \mu(\Omega_k(A/p)_{(0)})$; and by the domain case treated above, $\mu(\Omega_k(A/p)_{(0)}) = \dim A/p$. Q.E.D.

In particular, if $A = k[X_1, \ldots, X_n]/I$, where k is a perfect field and I is an unmixed radical ideal, then $\operatorname{rk}\Omega_k(A) = \dim A$.

An argument similar to the above may be used to derive conditions under which a zero may be inserted on the left of the exact sequence 31.2. Let $I \neq R$ be an ideal of $R = k[X_1, \ldots, X_n]$. Since $\Omega_k(R)$ is free of rk n, the exact sequence 31.2 of R/I-modules becomes

$$I/I^2 \xrightarrow{\phi} (R/I)^n \to \Omega_k(R/I) \to 0.$$

Let $K = \operatorname{Im}\phi$.

32.2. LEMMA: *Let k be a field, let $R = k[X_1, \ldots, X_n]$, and let $I \neq R$ be a radical ideal of R such that $(R/I)_q$ is separably generated over k for every minimal prime q of I. Then for any prime $p \supset I$, $\operatorname{rk} K_p = \operatorname{ht} I_p$.*

Proof: Since R/I is reduced, $(R/I)_q$ is a field for any minimal prime q of I; and therefore $\mu(K_q) = n - \mu(\Omega_k(R/I)_q)$. But, as we have seen above, $\mu(\Omega_k(R/I)_q) = \operatorname{coht} q$. Thus, $\mu(K_q) = n - \operatorname{coht} q = \operatorname{ht} q$. Since by definition $\operatorname{rk} K_p = \inf\{\mu(K_q) \mid q$ is a minimal prime of I and $q \subset p\}$, then $\operatorname{rk} K_p = \inf\{\operatorname{ht} q \mid q$ is a minimal prime of I and $q \subset p\} = \operatorname{ht} I_p$.

32.3. THEOREM: *In addition to the assumptions of* 32.2, *suppose* I *is locally generated by a regular sequence at every prime* $\supset I$. *Then* ϕ *is injective.*

Proof: It suffices to show ϕ_p is injective for every prime $p \supset I$. Since I_p is generated by an R_p-sequence of length t, $(I/I^2)_p$ is $(R/I)_p$-free of rk t by 28.1. Since K_p is a homomorphic image of $(I/I^2)_p$, then $\mu(K_p) \leqslant t$. On the other hand, $\operatorname{rk} K_p = \operatorname{ht} I_p = t$ by 32.2. Therefore $t = \operatorname{rk} K_p \leqslant \mu(K_p) \leqslant t$, so K_p is free of rk t by 15.1; and hence ϕ_p is injective.

32.4. COROLLARY: ϕ *is injective if* k *is a perfect field and* I *is a radical ideal of* $R = k[X_1, \ldots, X_n]$ *such that* I *is locally generated by a regular sequence at every prime* $\supset I$.

33. Fitting ideals (cf. [32, p. 145]) and the Jacobian ideal. Let M be a f.g. R-module. Given a presentation $0 \to K \to R^n \to M \to 0$, $n \geqslant 1$, we may regard the elements of K as n-tuples in R^n and may form the (usually infinitely long) matrix (K) whose rows are these n-tuples (in any order). The ith *Fitting ideal* $F_i(M)$ is defined to be the ideal of R generated by the $(n-i) \times (n-i)$ subdeterminants of (K) for $i = 0, \ldots, n-1$, and to be R for $i \geqslant n$. (Kaplansky begins his subscripts with 1 instead of 0.) Thus, $F_0 \subset F_1 \subset \cdots \subset F_n = F_{n+1} = \cdots = R$. It may be seen that these Fitting ideals do not depend on the particular presentation for M. Moreover, if K is finitely generated, then one does not have to use the full matrix (K) but may use instead the matrix whose rows constitute the generators of K, possibly augmented by enough additional rows of zeros so as to have at least n rows. Finally, Fitting ideals localize well: $F_i(M_S) = F_i(M)R_S$ for any m.s. S of R.

33.1. *Properties*: (a) For any prime ideal p of R, $\mu(M_p)$ may be characterized as the least integer i such that $F_{i-1}(M) \subset p$ and $F_i(M) \not\subset p$. For if $\mu(M_p) = n \geqslant 1$ and we resolve M_p on n generators,

$$0 \to K \to R_p^n \to M_p \to 0,$$

then the elements of the matrix (K) are necessarily in pR_p; and therefore $F_{n-1}(M_p) \subset pR_p$ and $F_n(M_p) = R_p$. Since $F_i(M)_p = F_i(M_p)$, we then have $F_{n-1}(M) \subset p$ and $F_n(M) \not\subset p$. On the other hand, if $\mu(M_p) = 0$, then $M_p = 0$ and $R_p = F_0(M_p) = F_0(M)_p$; so $F_0(M) \not\subset p$.

(b) $F_0(M) = R$ (if and) only if $M = 0$. For, $F_0(M) \not\subset p$ implies $\mu(M_p) = 0$ by (a); and $M_p = 0$ for every prime p of R implies $M = 0$.

(c) M_p is free of rk n if and only if $F_{n-1}(M_p) = 0$ and $F_n(M_p) = R_p$. For, $F_{n-1}(M_p) = 0 \Leftrightarrow (K)$ is the 0-matrix $\Leftrightarrow K = 0$.

(d) Lower and upper rk: There are two distinguished subscripts $\underline{r}(M)$ and $\bar{r}(M)$ that occur in the sequence of Fitting ideals for $M: 0 = F_0 = \cdots = F_{\underline{r}-1} < F_{\underline{r}} \subset \cdots \subset F_{\bar{r}-1} < F_{\bar{r}} = \cdots = R$. We shall call these numbers the *lower* and *upper* rks of M, respectively. Note that by the characterization of $\mu(M_p)$ given in (a) for a reduced ring R, $\underline{r}(M) = \inf\{\mu(M_p) \mid p \in \operatorname{Spec} R\} = \operatorname{rk} M$; and, similarly, for any R, $\bar{r}(M) = \sup\{\mu(M_p) \mid p \in \operatorname{Spec} R\}$.

Now let k be a field and $A = k[x_1, \ldots, x_n]$ be a f.g. k-algebra. We have seen in §31 that a presentation $A = k[X_1, \ldots, X_n]/I$ gives rise to a corresponding presentation for the module of differentials $\Omega = \Omega_k(A)$, $0 \to K \to A^n \to \Omega \to 0$, and that if $f_1, \ldots, f_m$ is a generating set for the ideal I, then the matrix (K) with respect to this generating set is just $\mathcal{J}(f_i; x) = (\partial f_i / \partial x_j)$.

By the definition of $\underline{r} = \underline{r}(\Omega)$ as the subscript of the first non-zero Fitting ideal of Ω, the rank of the matrix $\mathcal{J}(f_i; x)$ is $n - \underline{r}$, which, incidentally, shows rank $\mathcal{J}(f_i; x)$ is independent of the choice of generating set $\{f_i\}$ for I. The ideal $F_{\underline{r}}(\Omega)$, generated by the $(n - \underline{r}) \times (n - \underline{r})$ subdeterminants of $\mathcal{J}(f_i; x)$, is called the *Jacobian ideal* of A; it depends only on A and will be denoted $\mathcal{J}(A)$.

Let p be a prime ideal of A. In analogy with the notion of geometrically simple point introduced in §5, we shall call the local ring A_p *geometrically regular* if rank $\mathcal{J}(f_i; x)$ does not decrease when this matrix is reduced mod p, or equivalently, if $\mathcal{J}(A) \not\subset p$. (Note: The analogy is exact only if k is a perfect field and I is an unmixed radical ideal, since then rank $\mathcal{J}(f_i; x) = \operatorname{ht} I$ by 32.2; to avoid this technicality, we shall further assume at this point that A is of this form.) By 33.1, A_p is geometrically regular if and only if Ω_p is free of rk $\underline{r}(\Omega)$. To sum up,

33.2. PROPOSITION: *Let $A = k[X_1, \ldots, X_n]/I$, k a perfect field and I an unmixed radical ideal, and let p be a prime ideal of A. Then the following are equivalent:*

(i) *A_p is geometrically regular.*
(ii) $\mathcal{J}(A) \not\subset p$.
(iii) *Ω_p is free of* rk $\underline{r}(\Omega)$.

If we define A to be geometrically regular if every localization A_p is, then we have the following global form of this proposition.

33.3. COROLLARY: *For A as in 33.2, the following are equivalent:*

(i) *A is geometrically regular.*
(ii) $\mathcal{J}(A) = A$.
(iii) *Ω is projective of constant* rk $\underline{r}(\Omega)$.

Recall from 33.1 that for a reduced A, $\underline{r}(\Omega) = \operatorname{rk} \Omega$; so (iii) is equivalent to: (iii′) Ω is projective of constant rk.

Finally, taking into account our previous observation that geometric regularity and regularity coincide over a perfect field (cf. §5), we can combine 33.3 with 32.4 to obtain the theorem needed in §30:

33.4. THEOREM: *Let $R = k[X_1, \ldots, X_n]$ with k a perfect field, and let $I \neq R$ be an ideal of R. If I is unmixed and R/I is regular, then $(I/I^2) \oplus \Omega_k(R/I) \cong (R/I)^n$.*

REMARKS: Let A be a reduced f.g. k-algebra, k a field; let p be a prime ideal of A; and let $\mathcal{O}, m$ denote the local ring A_p, pA_p.

(a) The following theorem of Kunz [40, p. 177] clarifies the relationship between the notions of regular and geometrically regular: Suppose A is a domain and $\Omega_{k^t}(A)$ is finitely generated, where t is the characteristic of k. Then $\mathfrak{J}_{k^t}(A) \not\subset p$ if and only if A_p is regular.

We have seen that $\mathfrak{J}_k(A) \not\subset p$ if and only if A_p is geometrically regular; so, in particular, the two notions coincide if k is perfect. (It is illuminating to compute the ideals $\mathfrak{J}_k(A)$ and $\mathfrak{J}_{k^t}(A)$ for the example of §5.)

(b) If $\mathcal{O}, m$ and A are one-dimensional and k is perfect, it may be seen directly that $\mathcal{O}$ regular implies $\Omega = \Omega_k(\mathcal{O})$ is rk 1 free:

Since $\mathcal{O}/m$ is separably algebraic over k, for any $a \in \mathcal{O}$ there exists a monic $f(X) \in k[X]$ such that $f(a) \in m = (t)$ and $\partial f/\partial a \notin m$. But then $df(a) = (\partial f/\partial a)da \in \langle dt \rangle + t\Omega$, and hence $da \in \langle dt \rangle + t\Omega$. Thus, $\Omega \subset \langle dt \rangle + t\Omega$, and therefore by Nakayama's lemma, $\Omega = \langle dt \rangle$. To conclude, it remains to verify that dt is not annihilated by a non-zero element of $\mathcal{O}$ or, equivalently, that $\Omega_q \neq 0$ for every ht 0 prime q of $\mathcal{O}$. But $\Omega_q \cong \Omega_k(\mathcal{O}_q)$, and $\Omega_k(\mathcal{O}_q) \neq 0$ by 31.1 since $\mathcal{O}_q$, the quotient field of $\mathcal{O}/q$, is not algebraic over k.

(c) There is a comparable theorem to 33.2 which characterizes complete intersection local rings.

THEOREM (Lipman [43]): *Suppose k is perfect, and let $\Omega = \Omega_k(\mathcal{O})$. Then $\mathfrak{J}_k(\mathcal{O})$ is a principal regular ideal if and only if $d(\Omega) \leqslant 1$ and Ω/Ω^T is free; and these conditions imply $\mathcal{O}$ is a complete intersection ring. Here $\Omega^T = \{w \in \Omega \mid$ there exists a regular $a \in \mathcal{O}$ such that $aw = 0\}$.*

There are two parts to the proof: one is the theorem of Ferrand and Vasconcelos which asserts that $\mathcal{O}$ is a complete intersection if and only if $d(\Omega) \leqslant 1$, and which will be proved in §35; and the other is the theorem of Lipman that for an arbitrary quasi-local ring $\mathcal{O}$ and a f.g. $\mathcal{O}$-module M, $F_{\underline{r}}(M)$ is a principal regular ideal if and only if $d(M) \leqslant 1$ and M/M^T is free of rk $\underline{r}$ (cf. also [7, p. 331]).

(d) It is unknown if, for k of characteristic 0, the dual $\mathrm{Hom}_{\mathcal{O}}(\Omega_k(\mathcal{O}), \mathcal{O})$ is free implies $\mathcal{O}$ is regular ("Conjecture of Zariski-Lipman" [42]; see also [37], [63], [29]).

Similarly, it is unknown if, for k perfect and A and $\mathcal{O}$ one-dimensional domains, $\Omega_k(\mathcal{O})$ is torsion-free implies $\Omega_k(\mathcal{O})$ is free [7, p. 346]. See also [42, p. 896].

34. Application to rational curves. There is a large class of non-singular irreducible curves which are known classically to have cyclic module of differentials and which are therefore complete intersections by 30.3, namely, the curves of genus 0 or 1 (cf. [65]). This is elementary for curves of genus 0, as we shall now show, provided one begins with the characterization that a non-singular irreducible curve has genus 0 if and only if the quotient field of its affine ring is a simple transcendental extension of the ground field k; irreducible curves with this property are known as *rational curves*.

PROPOSITION: *Let k be an algebraically closed field and $k(t)$ be a simple transcendental extension of k. Then for any integrally closed domain D having quotient field $k(t)$ and such that $k \subset D \subset k(t)$, D is a* PID *and $\Omega_k(D)$ is* rk 1 *free.*

Proof: The valuation rings of $k(t)$ containing k are the $1/t$-adic and the $(t-\alpha)$-adic, $\alpha \in k$. Since D is integrally closed, it is an intersection of such valuation rings; and since k is the intersection of all such valuation rings, at least one of them is excluded from this representation for D. If the p-adic is excluded, where $p = 1/t$ or $t - \alpha$, then $1/p$ is in all the other valuation rings and hence is in D. Thus, by replacing t by $1/p$, we may assume $k[t] \subset D$.

It now suffices to see $D = k[t]_S$, where $S = \{f \in k[t] \mid 1/f \in D\}$. For then every ideal of D is extended from the PID $k[t]$ and consequently must be principal, and $\Omega_k(D) = \Omega_k(k[t])_S$ is rk 1 free because $\Omega_k(k[t])$ is. Thus, let $0 \neq a \in D$ and write $a = f/g$, $f, g \in k[t]$. Since $k[t]$ is a PID, we may assume $(f, g) = (1)$; and therefore $1 = rf + sg$, $r, s \in k[t]$. Then $1/g = ra + s \in D$. Q.E.D.

REMARK: The above proof also shows that any non-singular irreducible curve of genus 0 is isomorphic to a plane curve; for if $D = k[t]_S$ is a finitely generated extension of k, then $D = k[t, 1/s]$ for some $s \in S$. Sathaye [62] has proved that any non-singular irreducible curve of genus 0 or 1 is isomorphic to a plane curve; and he also gives an example of a non-singular irreducible curve of genus > 1 which is a complete intersection but is not isomorphic to a plane curve.

35. Complete intersections and resolutions of Ω. We prove in §35 two theorems relating complete intersections to the module of differentials. One is the theorem of Ferrand and Vasconcelos which asserts that a variety $\mathcal{V}$ is locally a c.i. if and only if the module of differentials Ω of its affine ring has projective dim $\leqslant 1$, and the other is the theorem of Mohan Kumar which asserts that $\mathcal{V}$ is isomorphic to a c.i. if and only if Ω has a f.g. free resolution of length $\leqslant 1$.

35.1. LEMMA [70]: *Let $I \neq R$ be an ideal of a local ring R, and suppose*

(i) *$d(I) < \infty$, and*

(ii) *there exists $t \geqslant 1$ and an R/I-module K such that $(R/I)^t \oplus K \cong I/I^2$. Then $G(I) \geqslant t$; and if $K \neq 0$, then $G(I) > t$.*

Proof: Condition (ii) may be rephrased: (ii′) there exist $a_1, \ldots, a_t \in I$ and an ideal L of R such that $I^2 \subset L$, $I = (a_1, \ldots, a_t, L)$, and $\sum r_i a_i \in L$ implies $r_i \in I$, $i = 1, \ldots, t$. Moreover, $K \neq 0$ if and only if $L \neq I^2$. The lemma will be proved by showing that there exists an R-sequence $b_1, \ldots, b_t$ such that $b_i \equiv a_i \bmod I^2$ and that if $L \neq I^2$, then there exists $c \in I$ such that $b_1, \ldots, b_t, c$ is also an R-sequence.

The hypothesis $d(I) < \infty$ implies either I contains a regular element z or $I = 0$ [32, p. 148], and the latter possibility is ruled out by (ii′). By adding a suitable multiple of z^2 to a_1, we may assume a_1 is regular. Now pass to the ideal $I^* = I/(a_1)$ in $R^* = R/(a_1)$, and consider the exact sequence of R^*-modules

$$0 \to (a_1)/a_1 I \to I/a_1 I \to I/(a_1) \to 0.$$

By [32, p. 124], $d_R(I) < \infty$ implies $d_{R^*}(I/a_1I) < \infty$; and since a summand of a module of finite projective dim again has finite projective dim, one may conclude that $d_{R^*}(I^*) < \infty$ after first noting that (ii′) implies the above sequence splits.

Now suppose $t = 1$. Then $d_{R^*}(I^*) < \infty$ implies as before that I^* contains a regular element c^* or $I^* = 0$. If $I^* = 0$, then $I = (a_1)$; and hence by (ii′), $L = I^2$. On the other hand, if $I^* \neq 0$, then a_1, c is the required R-sequence, where c denotes any pre-image in R for c^*.

To conclude the proof, note that condition (ii′) carries down to R^* and apply induction on t.

35.2. COROLLARY: *Let $I \neq R$ be an ideal of a local ring R, and suppose $d(I) < \infty$ and there exists a surjective R/I-module homomorphism $\phi: I/I^2 \to (R/I)^t$, for some $t \geqslant 1$. Then $G(I) \geqslant t$, and the following are equivalent:*

(i) $G(I) = t$.

(ii) *ϕ is an isomorphism.*

(iii) *I is generated by an R-sequence of length t.*

Proof: The assertions $G(I) \geqslant t$ and (i)$\Rightarrow$(ii) are just 35.1; the implication (ii)$\Rightarrow$(iii) follows by noting that $\mu(I) = t$ by Nakayama's lemma, since then $t \leqslant G(I) \leqslant \mu(I) = t$; and (iii)$\Rightarrow$(i) is immediate. Q.E.D.

For the remainder of §35, *assume $R = k[Z_1, \ldots, Z_n]$, k a perfect field and the Z_i indeterminates, and $I \neq R$ is a radical ideal of R.* Note that if $I \neq (0)$, then $I/I^2 \neq (0)$, for it is an easy exercise [10, p. 172] to see that a f.g. idempotent ideal is generated by an idempotent.

35.3. THEOREM (Ferrand [20], Vasconcelos [70]): *I is locally generated by a regular sequence at every prime $\supset I$ if and only if $d(\Omega_k(R/I)) \leqslant 1$.*

Proof: $\Rightarrow$: The sequence of R/I-modules $0 \to I/I^2 \to (R/I)^n \to \Omega_k(R/I) \to 0$ is exact by 32.4, and I/I^2 is R/I-projective by 28.1.

$\Leftarrow$: Consider the fundamental exact sequence $I/I^2 \xrightarrow{\phi} (R/I)^n \to \Omega_k(R/I) \to 0$, and let $K = \operatorname{Im}\phi$. Then $d(\Omega_k(R/I)) \leqslant 1$ implies K is projective, and hence K_p is free for any prime $p \supset I$. But $G(I_p) \geqslant \operatorname{rk} K_p = \operatorname{ht} I_p$, the inequality by 35.2 and the equality by 32.2. Since $G(I_p) \leqslant \operatorname{ht} I_p$ is valid in general, then $G(I_p) = \operatorname{rk} K_p$; and therefore another application of 35.2 concludes the proof. Q.E.D.

In defining the notion of c.i. local ring in §24, we have seen that if a local ring $\mathcal{O}$ can be written as a regular local ring modulo a c.i. ideal, then whenever $\mathcal{O}$ is written as a homomorphic image of a regular local ring, the kernel is necessarily a c.i. ideal. The corresponding global situation is not as simple, as the next theorem demonstrates; it and its corollary are due to Mohan Kumar [39].

35.4. THEOREM: *The following are equivalent:*

(i) *There exists an integer n' and a* c.i. *ideal I' in the polynomial ring $R' = k[X_1, \ldots, X_{n'}]$ such that $R'/I' \cong R/I$ (where $\cong$ denotes a k-algebra isomorphism).*

(ii) *There exist* f.g. *free R/I-modules F_0, F_1 such that $0 \to F_1 \to F_0 \to \Omega_k(R/I) \to 0$ is exact.*

(iii) *I/I^2 is a stably free R/I-module (Recall that a module M is called stably free if there exists a* f.g. *free module F such that $M \oplus F$ is* f.g. *free).*

Proof: (i)$\Rightarrow$(ii): I'/I'^2 is R'/I'-free by 28.1, and the sequence $0 \to I'/I'^2 \to (R'/I')^{n'} \to \Omega_k(R'/I') \to 0$ is exact by 32.4.

(ii)$\Rightarrow$(iii): By 35.3 I is locally generated by a regular sequence at primes $\supset I$. Therefore by 32.4 the sequence $0 \to I/I^2 \to (R/I)^n \to \Omega_k(R/I) \to 0$ is exact, and hence by Schanuel's lemma I/I^2 is stably free.

(iii)$\Rightarrow$(i): By (iii) there exists an integer $t \geqslant 0$ such that $(I/I^2) \oplus (R/I)^t \cong (R/I)^m$. Let I' be the ideal $(I, X_1, \ldots, X_t, Y)$ in $R[X_1, \ldots, X_t, Y]$. Claim: I' is a c.i. ideal. Let J be the ideal $(I, X_1, \ldots, X_t)$ in $R[X_1, \ldots, X_t]$. Since I/I^2 is R/I-projective of constant rk $m - t$, I is locally generated by a regular sequence of length $m - t$ at primes $\supset I$ by 35.2. Therefore $\operatorname{ht} I = m - t$, and hence $\operatorname{ht} I' \geqslant (m - t) + t + 1 = m + 1$, so it remains to show $\mu(I') \leqslant m + 1$.

Since any linear element $a_0 + a_1X_1 + \cdots + a_tX_t, a_i \in R$, of $R[X_1, \ldots, X_t]$ is in J (respectively, J^2) if and only if $a_0 \in I$ (respectively, $a_0 \in I^2$ and $a_1, \ldots, a_t \in I$), it follows that J/J^2 is isomorphic as an R/I-module to $(I/I^2) \oplus (R/I)^t$ and hence also to $(R/I)^m$. Thus, $\mu_{R/I}(J/J^2) = m$. Now $\mu(I') \leqslant m + 1$ follows from

LEMMA: *Let S be a ring, let J be a* f.g. *ideal of S, and let Y be a single indeterminate. If the S/J-module J/J^2 can be generated by m elements, then the ideal (J, Y) of $S[Y]$ can be generated by $m + 1$ elements.*

Proof: Choose $a_1, \ldots, a_m \in J$ such that $J = (a_1, \ldots, a_m) + J^2$. Then the ideal $J^* = J/(a_1, \ldots, a_m)$ in $S^* = S/(a_1, \ldots, a_m)$ is idempotent. Since a f.g. idempotent ideal is generated by an idempotent element, $J^* = (e^*)$ with $e^{*2} = e^*$. Then in $S^*[Y]$ we have $(J^*, Y) = (e^*, Y) = (e^* + (1 - e^*)Y)$; so (J^*, Y) is principal in $S^*[Y]$, and hence (J, Y) is generated by $m + 1$ elements. Q.E.D.

We have seen (30.4) that any non-singular curve which is isomorphic to a c.i. is itself a c.i. A corresponding result holds for any variety of $\dim n - 2$ in $\mathcal{C}^n$ (Question: Is there a common generalization?):

35.5. COROLLARY [39]: *If* $\operatorname{ht} I = 2$ *and there exists an integer $n' \geqslant 1$ and an ideal I' of $R' = k[X_1, \ldots, X_{n'}]$ such that $R/I \cong R'/I'$ (as k-algebras), then I' is a* c.i. *ideal implies I is a* c.i. *ideal.*

Proof: I/I^2 is a stably free R/I-module by 35.4, so $(I/I^2) \oplus (R/I)^{m-t} \cong (R/I)^m$ for some $m \geqslant t \geqslant 1$. Then I/I^2 is R/I-projective of constant $\operatorname{rk} t$, and hence I is locally generated by a regular sequence of length t at primes $\supset I$ by 35.2. Moreover, then $t = \operatorname{ht} I = 2$. By 30.1 $\operatorname{Ext}^1_R(I, R) \cong R/I$, and therefore by 13.2 I is a c.i. ideal. Q.E.D.

REMARK: Scheja–Storch [64] have proved some theorems closely related to those of §35; their results have an interesting setting that arises from their use of Noether normalization to replace the ground field k by a polynomial ring $k[t_1, \ldots, t_s] \subset R/I$ such that R/I is a finite module over $k[t_1, \ldots, t_s]$.

FURTHER READING

The bibliography includes a few references which have not been mentioned in the text proper and which I feel would provide interesting further reading. Here is a brief summary of the contents of these additional references.

(a) F.G. *projectives with free summands*: It is sometimes possible to make use of a weaker assumption than f.g. projectives of a certain rk being free in passing from a bound for $\mu_1(I)$ to one for $\mu(I)$ as described in section II. We have in mind the following condition:

Condition $(\mathbb{S}_r)$: Any f.g. projective R-module P of rk $\geqslant r$ has a non-zero free summand (i.e., $P \cong Q \oplus R$ for some Q).

Murthy-Swan [50] have studied the case of a 2-dim regular domain R which is a finitely generated extension of an algebraically closed field k; geometrically, R is the affine ring of a non-singular irreducible surface. They show that such an R satisfies $(\mathbb{S}_2)$ if it defines a ruled surface, i.e., if its quotient field K is a simple transcendental extension of a field intermediate to k and K, but not in general. Moreover, they prove that R satisfies $(\mathbb{S}_2)$ if and only if $\mu(m) \leqslant 2$ for every maximal ideal m of R.

Note also that for any ring R satisfying $(\mathbb{S}_2)$, $\mu(I) \leqslant 2$ for every invertible ideal I of R; for write $I \oplus I^{-1} \cong Q \oplus R$ and take $\bigwedge^2$ of both sides to get $R = II^{-1} \cong Q$, which implies $\mu(I) \leqslant 2$ since $Q \oplus R$ projects onto I. Murthy–Swan have proved that for R the affine ring of a non-singular surface the converse to this is false in general but is true if every maximal ideal m of R is of the form

$m = p_1 + p_2$, where p_i is a ht 1 prime (geometrically, every point of the surface should be a complete intersection, ideal-theoretically, of two irreducible curves on the surface).

A particular kind of ruled surface is that given by an affine ring of the form $D[X]$, D a Dedekind domain. For *any* ideal I of $D[X]$, with D an arbitrary Dedekind domain, an unpublished result, due independently to Murthy and Davis–Geramita, asserts that $\mu(I) \leqslant 2$ locally implies $\mu(I) \leqslant 2$; this may be regarded as establishing the Eisenbud-Evans conjecture [18] for ideals of such a ring. See also [16] for related results on the minimal number of generators of maximal ideals.

(b) *Curves defined by numerical monoids*: Herzog and Kunz (cf. [26]) have studied the complete intersection question for curves in $\mathfrak{A}^n$ having parametric equations of the form $X_1 = t^{s_1}, \ldots, X_n = t^{s_n}$, where $s_1, \ldots, s_n$ are integers > 0 with gcd $= 1$, i.e., the question of when the ht $n-1$ prime ideal p given by $0 \to p \to k[X_1, \ldots, X_n] \to k[t^{s_1}, \ldots, t^{s_n}] \to 0$ is generated by $n-1$ elements. The curve defined by p is an irreducible rational curve and may be seen to have only one singularity, at the origin. See also [11].

(c) *Recent developments*: Since this work was completed, a number of important results have appeared.

One of these is Mohan Kumar's proof of the full Eisenbud–Evans conjecture ["On two conjectures about polynomial rings," *Invent. Math.*, **46** (1978), 225–236]. His theorem yields as a corollary Forster's conjecture that any locally c.i. ideal of $k[X_1, \ldots, X_n]$, k a field, is generated by n elements. Sathaye ["On the Forster-Eisenbud-Evans conjecture," ibid., 211–224] has obtained similar results, but with restrictive hypotheses.

Murthy [49] has asked if every locally c.i. ideal I of $R = k[X_1, \ldots, X_n]$ with I/I^2 free over R/I is a c.i. M. Boratynski ["A note on the set-theoretical complete intersection ideals," *J. of Alg.*, 54 (1978), 1–5] has applied a theorem of Suslin to show that such an I is at least a set-theoretic c.i. Also, Mohan Kumar proves in the above work that if $\mu(I/I^2) \geqslant \operatorname{coht} I + 2$, then $\mu(I) = \mu(I/I^2)$.

R. Cowsik and M. Nori ["Affine curves in characteristic p are set theoretic complete intersections," *Invent. Math.*, **45** (1978), 111–114] have proved the theorem of their title.

The definitive exposition of the Serre problem is now T. Y. Lam's notes [*Serre's Conjecture*, Lecture Notes in Math., No. 635, Springer-Verlag, New York, 1978].

T. Matsuoka ["On almost complete intersections," *Manuscripta Math.*, **21** (1977), 329–340] has observed that the kernel of the map ϕ discussed in §32 is just the torsion submodule of I/I^2; thus, it can be seen that under the hypotheses of 32.2 one has an exact sequence of R/I-modules

$$0 \to (I/I^2)^T \to I/I^2 \xrightarrow{\phi} (R/I)^n \to \Omega_k(R/I) \to 0.$$

The question raised at the end of §4 has come to be known as "the prime sequence question." A counterexample has recently been given by R. Heitmann ["A negative answer to the prime sequence question," *Proc. Amer. Math. Soc.*, 77 (1979), 23–26], while some positive results have been obtained for affine domains over an infinite ground field by E. D. Davis ["Prime elements and prime sequences in polynomial rings," *Proc. Amer. Math. Soc.*, 72 (1978), 33–38] and N. V. Trung [Correspondence, 1978; cf. Dissertation, Martin-Luther-Univ., Halle, 1977].

REFERENCES

To keep the size of the bibliography in bounds, I have in a number of cases referred to a paper with an up-to-date list of references in preference to a primary source. Thus, while the following list omits many fundamental works, if the reader traces backward from it, he should get a fairly complete picture of what is known.

1. S. S. Abhyankar, "Algebraic space curves," *Séminaire Math. Sup.*, **43**, Montréal, 1971.
2. ——, *On Macaulay's Examples* (notes by A. Sathaye, Conference on Commutative Algebra), Lecture Notes in Math., no. 311, Springer-Verlag, New York, 1973, 1–16.
3. S. S. Abhyankar and A. M. Sathaye, *Geometric Theory of Algebraic Space Curves*, Lecture Notes in Math., no. 423, Springer-Verlag, New York, 1974.

4. A. Altman and S. Kleiman, *Introduction to Grothendieck Duality Theory*, Lecture Notes in Math., no. 146, Springer-Verlag, New York, 1970.
5. M. Artin, *Commutative Rings*, MIT Notes, 1966.
6. H. Bass and M. P. Murthy, "Grothendieck groups and Picard groups of abelian group rings," *Ann. of Math.* **86** (1967), 16–73.
7. R. Berger, "Differentialmoduln eindimensionaler lokaler Ringe," *Math. Z.*, **81** (1963), 326–354.
8. R. Berger and E. Kunz, "Über die Struktur der Differentialmoduln von diskreten Bewertungsringen," *Math. Z.*, **77** (1961), 314–338.
9. N. Bourbaki, *Algèbre Multilinéaire*, Éléments de Mathématique, vol. 3, Hermann, Paris, 1958.
10. ———, *Algèbre Commutative*, Éléments de Mathématique, vol. 27, Chaps. 1 and 2, Hermann, Paris, 1961.
11. H. Bresinsky, "Symmetric semigroups of integers generated by 4 elements," *Manuscripta Math.* (1976), 205–220.
12. W. Bruns, "Zur erzeugung von moduln," *Comm. in Alg.*, **4** (1976), 341–373.
13. ———, "Jede endliche freie Auflösung ist freie Auflösung eines von drei Elementen Erzeugten Ideals," *J. Algebra*, **39** (1976), 429–439.
14. I. S. Cohen, "On the structure and ideal theory of complete local rings," *Trans. Amer. Math. Soc.*, **59** (1946), 54–106.
15. E. D. Davis, "Further remarks on ideals of the principal class," *Pacific J. Math.*, **27** (1968), 49–51.
16. E. D. Davis and A. V. Geramita, "Efficient generation of maximal ideals in polynomial rings," *Trans. Amer. Math. Soc.*, **231** (1977), 497–505.
17. D. Eisenbud and E. G. Evans, "Generating modules efficiently: Theorems from algebraic K-theory," *J. Algebra*, **27** (1973), 278–305.
18. ———, "Three conjectures about modules over polynomial rings," *Springer Lecture Notes in Math.*, **311** (1973), 78–89.
19. ———, "Every algebraic set in n-space is the intersection of n hypersurfaces," *Invent. Math.*, **19** (1973), 107–112.
20. D. Ferrand, "Suite régulière et intersection complète," *C. R. Acad. Sci. Paris Ser. A*, **264** (1967), 427–428.
21. ———, *Les modules projectifs de type fini sur un anneau polynomes sur un corps sont libres* (d'après Quillen et Suslin), Séminaire Bourbaki, no. 484, 1975/76, pp. 1–20.
22. D. Ferrand and M. Raynaud, "Fibres formelles d'un anneau local noetherien," *Ann. Sci. Ecole Norm. Sup.*, **3** (1970), 295–311.
23. W.-D. Geyer, "On the number of equations which are necessary to describe an algebraic set in n-space," *Atas da 3ª Escola de Algebra*, Brasilia, 1976; IMPA Lecture Notes, 9, Rio de Janeiro, 1977, pp. 183–317.
24. W. Gröbner, *Moderne Algebraische Geometrie*, Springer-Verlag, Wien and Innsbruck, 1949.
25. R. Hartshorne, "Varieties of small codimension in projective space," *Bull. Amer. Math. Soc.*, **80** (1974), 1017–1032.

26. J. Herzog and E. Kunz, Die Wertehalbgruppe eines lokalen Rings der Dimension I, *S.-B. Heidelberger Akad. Wiss.* (1971), 27–43.
27. M. Hochster, "Constraints on systems of parameters," *Proc. Oklahoma Ring Theory Conference*, Marcel-Dekker, New York, 1974, pp. 121–161.
28. ———, "Big Cohen-Macaulay modules and algebras and embeddability in rings of Witt vectors," *Conference on Commutative Algebra—1975*, Queen's Papers in Pure and Applied Math., no. 42, Queen's University, Kingston, Ontario, 1975, pp. 107–195.
29. ———, "On the Zariski-Lipman conjecture for homogeneous complete intersections," *Proc. Amer. Math. Soc.*, **49** (1975), 261–262.
30. M. Hochster and J. L. Roberts, "Rings of invariants of reductive groups acting on regular rings are Cohen-Macaulay," *Advances in Math.*, **13** (1974), 115–175.
31. F. Ischebeck, "Gewisse additive Funktionen auf Modulkategorien," *Arch. Math.*, **22** (1971), 252–259.
32. I. Kaplansky, *Commutative Rings*, University of Chicago Press, Chicago, 1974.
34. ———, *Topics in Commutative Ring Theory*, University of Chicago Notes, 1974.
35. R. Kiehl and E. Kunz, "Vollständige Durchschnitte und p-Basen," *Arch. Math.*, **26** (1965), 348–362.
36. M. Kneser, "Über die Darstellung algebraischer Raumkurver als Durchschnitte von Flächen," *Arch. Math.*, **11** (1960), 157–158.
37. H. Krämer, "Eine Bemerkung zu einer Vermutung von Lipman," *Arch. Math.*, **20** (1969), 30–35.
38. ———, "Einige Anwendungen der G-Function von MacRae," *Arch. Math.*, **22** (1972), 479–490.
39. N. Mohan Kumar, "Complete intersections," manuscript, 1976.
40. E. Kunz, "Die Primidealteiler der Differenten in allgemeinen Ringen," *J. Reine Angew. Math.*, **204** (1960), 165–182.
41. S. Lefschetz, *Algebraic Geometry*, Princeton University Press, Princeton, N.J., 1953.
42. J. Lipman, "Free derivation modules on algebraic varieties," *Amer. J. Math.*, **87** (1965), 874–898.
43. ———, "On the Jacobian ideal of the module of differentials," *Proc. Amer. Math. Soc.*, **21** (1969), 422–426.
44. S. Mac Lane, *Homology*, Springer, Berlin, 1963.
45. R. MacRae, "On an application of the Fitting invariants," *J. Algebra*, **2** (1965), 153–169.
46. H. Matsumura, *Commutative Algebra*, Benjamin, New York, 1970.
47. T. T. Moh, "On the unboundedness of generators of prime ideals in power series rings of three variables," *J. Math. Soc. Japan*, **26** (1974), 722–734.
48. M. P. Murthy, "Generators for certain ideals in regular rings of dimension three," *Comment. Math. Helv.*, **47** (1972), 179–184.

49. ———, "Complete intersections," *Conference on Commutative Algebra—1975*, Queen's Papers in Pure and Applied Math., no. 42, Queen's University, Kingston, Ontario, 1975, 197–211.
50. M. P. Murthy and R. G. Swan, "Vector bundles over affine surfaces," *Invent. Math.*, **36** (1976), 125–165.
51. M. Noether, "Zur grundlegung der Theorie der algebraische Raumcurven," *Abhandlungen Akad. Wiss. Berlin* (1882), 1–120.
52. D. G. Northcott, *Lessons on Rings, Modules and Multiplicities*, Cambridge University Press, Cambridge, 1968.
53. J. Ohm, "Some counterexamples related to integral closure in $D[[X]]$," *Trans. Amer. Math. Soc.*, **122** (1966), 321–333.
54. ———, "An axiomatic approach to homological dimension," *Math. Scand.*, **37** (1975), 197–222.
55. J. Ohm and R. L. Pendleton, "Rings with noetherian spectrum," *Duke Math. J.*, **35** (1968), 631–640.
56. O. Perron, "Studien über den Vielfachheitsbegriff und den Bézoutschen Satz," *Math. Z.*, **19** (1944), 654–679.
57. C. Peskine and L. Szpiro, "Liaison des variétés algébriques, I," *Invent. Math.*, **26** (1974), 271–302.
58. D. Quillen, "Projective modules over polynomial rings," *Invent. Math.*, **36** (1976), 167–171.
59. D. E. Rush, "The G-function of MacRae," *Canad. J. Math.*, **26** (1974), 854–865.
60. J. Sally, *Numbers of generators of ideals in local rings*, Pure Appl. Math. Series, no. 35, Dekker, New York, 1978.
61. J. Sally and W. Vasconcelos, "Stable rings," *J. Pure Applied Alg.*, **4** (1974), 319–336.
62. A. Sathaye, "On planar curves," *Amer. J. Math.*, 99 (1977), 1105–1135.
63. G. Scheja and U. Storch, "Differentielle Eigenschaften der Lokalisierungen analytischer Algebren," *Math. Ann*, **197** (1972), 137–170.
64. ———, "Quasi-Frobenius-Algebren und lokale vollständige durchschnitte," *Manuscripta Math.*, **19** (1976), 75–104; addendum, ibid., **20** (1977), 99–100.
65. A. Seidenberg, *Elements of the Theory of Algebraic Curves*, Addison-Wesley, Reading, Mass., 1968.
66. J. P. Serre, *Modules projectifs et espaces fibrés à fibres vectorielles*, Séminaire Dubreil-Pisot, no. 23, 1957/58.
67. ———, *Sur les modules projectifs*, Séminaire Dubreil-Pisot, no. 2, 1960/61.
68. F. Severi, "Il concetto generale di molteplicità . . . ," *Memorie Scelte*, vol. 1, Cremonese, Rome, 1950, pp. 327–390.
69. U. Storch, "Bemerkung zu einem satz von Kneser," *Arch. der Math.*, **23** (1972), 403–404.
70. W. Vasconcelos, "A note on normality and the module of differentials," *Math. Z.*, **105** (1968), 291–293.
71. ———, "On projective modules of finite rank," *Proc. Amer. Math. Soc.*, **22** (1969), 430–433.

72. ———, *Divisor Theory in Module Categories*, Math. Studies, no. 14, North-Holland/American Elsevier, New York, 1974.
73. A. Weil, *Foundations of Algebraic Geometry*, Amer. Math. Soc. Colloquium Publ., no. 29, Providence, R.I., 1962.
74. O. Zariski, "The concept of a simple point of an abstract algebraic variety," *Trans. Amer. Math. Soc.*, **62** (1947), 1–52.
75. O. Zariski and P. Samuel, *Commutative Algebra*, vol. 2, Van Nostrand, Princeton, N.J., 1960.

CHASLES'S ENUMERATIVE THEORY OF CONICS: A HISTORICAL INTRODUCTION

Steven L. Kleiman

Since antiquity, people have been intrigued by problems like the "problem of Apollonius" (circa 200 B.C.), construct a circle tangent to 3 given ones. Jacob Steiner [39, §II, 6 (1848)] pointed out that when the circles are replaced by arbitrary conics in the complex projective plane $\mathbf{P}^2$, these problems become more involved. It takes not 3 but 5 conditions to determine conics. Five conditions rarely determine a single conic, so the first task is to find out how many there are. Steiner discussed briefly a typical example, which has become famous. He said he believed that

The research for this paper was partially supported by Danish NSRC No. 511-5684. Many thanks go out for the generous hospitality I received as a guest at Copenhagen University and Aarhus University. Special thanks also go to Robert Speiser; joint work with him on a treatment of the mathematical justification of Chasles's theory gave rise to my delving into the history discussed here. Special thanks also go to the editor, Abraham Seidenberg; he read the manuscript carefully and made a number of suggestions for improving the exposition, which have been incorporated.

$6^5 = 7776$ conics are, in general, tangent to 5 others. He said little about how he found the number; he said only that 6 conics pass through 4 points and are tangent to 1 conic, 6^2 conics pass through 3 points and are tangent to 2 conics, and continuing in this manner leads to 6^5. (Similar stepwise procedures had been employed previously by William Braikenridge [2 (1733)], according to Zeuthen-Pieri [53, 20 (1915)].)

Johann Bischoff [1, especially p. 172 (1859)] looked at matters essentially in the following way. The conics are parametrized by the points of $\mathbf{P}^5$, via the coefficients of their defining equations. The conics tangent to a given conic form a hypersurface of degree 6. The conics passing through r points form a linear space of dimension $(5 - r)$. Hence there are 6^{5-r} conics passing through r points and tangent to $5 - r$ conics by Bezout's theorem; in particular there are $6^5 = 7776$ conics tangent to 5 conics. In fact, Bischoff proved that the hypersurface of curves of degree m tangent to a curve of degree n has degree $n(n + 2m - 3)$. This result had been stated without proof earlier by Steiner [40, p. 6 (1854)]. Bischoff's proof is basically all right, and the result became associated with his name. (Bischoff, however, did not use the geometric language of hypersurfaces in a parametrizing projective space.)

Ernest de Jonquières [10 (1861)] (a naval officer and mathematician) introduced an important new point of view. He found that in many cases the number of conics (in fact, curves of arbitrary degree) in a 1-parameter family that satisfy a simple condition is equal to $\alpha\mu$, where α depends only on the condition and μ denotes the number of conics in the family that pass through a given point. Now, α is just the degree of the hypersurface of conics satisfying the condition, and μ is just the degree of the curve parametrizing the family; hence the number $\alpha\mu$ results once again from Bezout's theorem. De Jonquières built up step by step to find the number of members of a several parameter family satisfying several conditions.

It was soon found that the stepwise procedure could lead to wrong results. For example, the number of conics tangent to 5 general lines is 1 not 32, although the hypersurface of conics tangent to a line does have degree 2. The true reason for this failure was discovered, according to Zeuthen-Pieri [53, 20 (1915)], by Luigi Cremona [8 (1864)]: the double lines in a family of conics were being counted as tangent to every curve. Thus the hypersurface of conics tangent to a given curve always contains the variety V in $\mathbf{P}^5$ parametrizing the double lines; V is now known as the Veronese surface. Five such hypersurfaces can never intersect in a finite number of points. In particular, the number 7776 has no enumerative significance.

Michel Chasles figured the expression $\alpha\mu$ should be replaced by one involving a conic C and the dual conic $\check{C}$ symmetrically. (He discusses the formation of his ideas at the bottom of p. 820 in [5 (1866)]. He was around 70 at the time but still a tireless worker.) The dual $\check{C}$ is the set of tangent lines to C; it is viewed as a point-set in the dual projective plane $\check{\mathbf{P}}^2$, whose points represent the lines in $\mathbf{P}^2$. For example, consider the conic $C: x^2 + y^2 + z^2 = 0$. At a point (a, b, c) of C the tangent line is $l: ax + by + cz = 0$, and l is represented by the triple (a, b, c) of coefficients. Hence $\check{C}$ is the locus in $\check{\mathbf{P}}^2$ of the equation $a^2 + b^2 + c^2 = 0$. In general, if C is represented by a symmetric matrix M, then $\check{C}$ is represented by M^{-1}. Indeed, change coordinates so that M becomes the identity. Then the assertion holds by the example. (Notice that if the coordinate change is given by a matrix N, then M changes to ${}^tNMN$ and a matrix $\check{M}$ representing $\check{C}$ changes to $N^{-1}\check{M}({}^tN^{-1})$.)

Chasles [4, p. 215 (1864)] replaced the expression $\alpha\mu$ by

$$\alpha\mu + \beta\nu.$$

Here μ is, as before, the number of conics in the 1-parameter family passing through a given point, and ν is, dually, the number tangent to a given line. Again α and β depend only on the given condition. Chasles [3, p. 298 (1864)] called μ and ν the *characteristics* of the family. The expression $\alpha\mu + \beta\nu$, with μ and ν viewed as variables, he [4, p. 348 (1864)] called the *module* of the condition. The numbers α, β have become known as the *characteristics* of the conditions, in keeping with Hermann Schubert's [29 (1879)] greater conception of the notion.

Some hard feelings between de Jonquières and Chasles came out in print [11 (1866)], [5 (1866)]. In short, de Jonquières felt Chasles should have given him some credit for inspiring him. Chasles answered that he was surprised at this. After all, de Jonquières's formula $\alpha\mu$ is wrong; $\alpha\mu + \beta\nu$ is right. So to cite de Jonquières would be doing him no favor.

Chasles [4, p. 216 (1864)] gave the following expression for the number $N(ZZ'Z''Z'''Z^{\text{iv}})$ of conics satisfying 5 independent simple conditions $Z, Z', Z'', Z''', Z^{\text{iv}}$:

$$\alpha\alpha'\alpha''\alpha'''\alpha^{\text{iv}} + 2\sum\alpha\alpha'\alpha''\alpha'''\beta^{\text{iv}} + 4\sum\alpha\alpha'\alpha''\beta'''\beta^{\text{iv}}$$
$$+ 4\sum\alpha\alpha'\beta''\beta'''\beta^{\text{iv}} + 2\sum\alpha\beta'\beta''\beta'''\beta^{\text{iv}}$$
$$+ \beta\beta'\beta''\beta'''\beta^{\text{iv}} \qquad (*)$$

where the α's and β's are the characteristics of the conditions.

Chasles derived (*) as follows. The number $N(ZZ'Z''Z'''Z^{\text{iv}})$ is given by the expression $\alpha^{\text{iv}}\mu + \beta^{\text{iv}}\nu$, where μ and ν are the characteristics of the 1-parameter family $(ZZ'Z''Z''')$ defined by the indicated conditions. By definition μ and ν are the numbers,

$$N(ZZ'Z''Z'''P) \quad \text{and} \quad N(ZZ'Z''Z'''L),$$

where P denotes the condition "to pass through a point" and L the condition "to be tangent to a line." The first number, μ, can be found similarly using the characteristics of the family $(ZZ'Z''P)$. These characteristics are the numbers,

$$N(ZZ'Z''PP) \quad \text{and} \quad N(ZZ'Z''PL).$$

Continuing this way leads ultimately to the 5 families defined entirely by points and lines. Chasles called these 5 families the *elementary systems*. (The 5 are discussed with equations and illustrations in Clebsch-Lindemann [25, pp. 393–397 (1876)].) Chasles remarked that the coefficient of the rth term in (*) is the number of conics passing through r points and tangent to $5 - r$ lines.

Halphen [13 (1873)], [14 (1873)] observed that (*) can be factored into the formal product,

$$(\alpha\mu + \beta\nu)(\alpha'\mu + \beta'\nu)(\alpha''\mu + \beta''\nu)(\alpha'''\mu + \beta'''\nu)(\alpha^{\mathrm{iv}}\mu + \beta^{\mathrm{iv}}\nu), \tag{*'}$$

of the modules of the conditions, provided that when the indicated multiplication is carried out, the term $\mu^r\nu^s$ is replaced by the number of conics passing through r points and tangent to s lines. Seven years earlier Eugène Prouhet [26, p. 202 (1866)] published a similar factorization, but Prouhet's observation seems to have led no further. (Prouhet was a répétiteur at the École Polytechnique and an editor of its journal. He was properly impressed by Chasles's theory and explained it [26 (1866)], [27 (1866)] in the journal.) On the other hand Halphen made a point of the factorization, and his note [14 (1873)] inspired Schubert (see Lit. 3, p. 333 in [29 (1879)]. Schubert developed this symbolic multiplication into a powerful calculus and enriched it with numerous examples. His work became the subject of David Hilbert's 15th problem [21, p. 464 (1902)] (see also [23 (1974)]); the problem calls for making Schubert's theory—which includes Chasles's—rigorous. The continuing effort to do so has given rise to much important mathematics.

Sometimes Bischoff's result as well as other results were used in conjunction with Chasles's theory. For example (see Clebsch-Lindemann [25, p. 401 (1876)]) the characteristics α, β of the condition of tangency to a given conic satisfy

$$6 = \mu^4(\alpha\mu + \beta\nu) = \alpha + 2\beta$$

because by Bischoff's result there are 6 conics through 4 points and tangent to a conic. By symmetry α and β are equal. Hence we obtain $\alpha = \beta = 2$.

The number of conics tangent to 5 others can now be computed using (*'):

$$\begin{aligned}(2\mu + 2\nu)^5 &= 32\left(1 + 2\binom{5}{1} + 4\binom{5}{2} + 4\binom{5}{3} + 2\binom{5}{4} + 1\right)\\ &= 32(102) = 3264.\end{aligned}$$

This number is correct and Chasles [3, p. 223 (1864)] was the first to publish it. It, along with more general results of the same sort, had been found earlier by de Jonquières (see footnote 2 on p. 315 of [12 (1866)]) using degeneration techniques; but de Jonquières left it unpublished, because it disagreed with the number 7776 announced by the much respected Steiner. Steiner himself had tried to get a feeling for the number 7776 by degenerating the given conics. Théodore Berner [Diss., Berlin (1865)] obtained 3264 the same way.

Francesco Severi [35, footnote p. 116 (1902)] obtained 3264 using a formula he found for the number of isolated intersections in $\mathbf{P}^n$ of n general hypersurfaces passing doubly through a common surface. Severi's method can be made rigorous. Surprisingly, doing so involves much the same formal considerations (centered around the blowup along the surface) as making Chasles's method rigorous. Moreover, Severi's method is asymmetric with respect to duality, and it misses out on something beautiful, the geometry of complete conics.

Chasles [3 (1864)], [4 (1864)], [6, (1871)] supported his theory with around 200 different examples. Each was worked out directly and the number of solutions then expressed in the form $\alpha\mu + \beta\nu$. Chasles did not pretend he had a proof that the form could always be achieved, although he did feel it could. Such a proof has been sought by many others, including Alfred Clebsch [7 (1873)] (who wrote it at the untimely end of his life), George-Henri Halphen [13 (1873)], [14 (1873)] (who to his regret lost the honor of giving the first proof), Ferdinand Lindemann [25, pp. 397–399 (1876)] (who added it to his write-up of Clebsch's book), Schubert and (his 17-year-old gymnasium student) Adolf Hurwitz [33 (1876)], Eduard Study [41 (1886)], [43 (1892)], [45 (1892)], Severi [37, §10 (1916)], [38 (1940)], and Bartel van der Waerden [47 (1938)]. Clebsch, Lindemann, and Study based their proofs on the invariant theory of ternary quadratic forms; Schubert and Hurwitz based theirs on the correspondence principle (on which Chasles based most of his examples (see [3, pp. 1172–1175 (1864)] and [27 (1866)]); Severi and van der Waerden based their proofs on algebraic intersection theory. On the basis of this work and of advances in algebraic geometry in general, Chasles's theory can now be readily understood and rigorously justified.

The justification of Chasles's theory begins with the notion of complete conic. A *complete conic* consists of a conic in $\mathbf{P}^2$ and one in $\check{\mathbf{P}}^2$ so related that the pair is the limit of a varying smooth conic and its dual. A complete conic may be viewed as a conic in $\mathbf{P}^2$ plus the structure of an envelope of tangent lines. For example, the conic

$$\frac{x^2}{b^2+c^2}+\frac{y^2}{b^2}=1$$

has the conic $(b^2+c^2)u^2+b^2v^2=1$ as its dual. When the first conic is degenerated into the double line $y^2=0$, by letting b tend to 0, the second conic degenerates into the line-pair $c^2u^2=1$. Note that the first conic degenerates in a family of confocal ellipses and that the two foci correspond to the two lines of the line-pair.

The name "vollständiger Kegelschnitt" (complete conic) was given by van der Waerden [47, p. 647 (1938)]. The concept is inherent in the work of all the previous authors up to and including Steiner [39 (1848)] (see footnote *, p. 189, for example), although they used the term "conic." Study [41 (1886)] in particular developed the notion of complete conic formally (although he says in footnote **, p. 71 that his results are mostly already in an 1884 article of Giuseppe Veronese and an 1885 article of Corrado Segre).

The variety B of complete conics is by definition the closure in $\mathbf{P}^5\times\check{\mathbf{P}}^5$ of the locus of pairs consisting of a point representing a smooth conic C and a point representing its dual $\check{C}$. If C is represented by a matrix M, then $\check{C}$ is represented by M^{-1}—so also by Λ^2M, the matrix of signed 2×2-minors of M, which differs from M^{-1} by a scalar multiple. Hence the duality correspondence extends over $\mathbf{P}^5$ up to the Veronese surface V, which is (ideal-theoretically) the locus of zeros of the 2×2-minors, and B is the closure of the graph of the extended correspondence. Therefore the projection $p:B\to\mathbf{P}^5$ identifies B as the blowup of $\mathbf{P}^5$ along V. Hence B is a smooth, irreducible variety of dimension

5, and the exceptional locus $p^{-1}V$ is a smooth irreducible variety of dimension 4. Of course the situation is symmetric in the two projections, $p: B \to \mathbf{P}^5$ and $q: B \to \check{\mathbf{P}}^5$.

Fix 5 smooth conics $C_1, \ldots, C_5$. The complete conics tangent to C_i (or, what is the same, to $(C_i, \check{C}_i)$) are parametrized by an irreducible subvariety Z_i of B with codimension 1. So the complete conics tangent to $C_1, \ldots, C_5$ are represented by the points of intersection $\cap Z_i$. Hence the number of them, if it is finite and weighted with the appropriate multiplicities, can be computed by forming the product $\pi \mathbf{z}_i$, where $\mathbf{z}_i$ is the class of Z_i in the intersection ring $A(B)$ of cycles modulo numerical equivalence. (Cycles are linear combinations of subvarieties. Two are numerically equivalent if for every third cycle intersecting the two properly the number of isolated points, weighted with the appropriate multiplicities, in each intersection is the same.) Numerical equivalence is the most natural equivalence relation to take for solving enumerative problems in which the figures are conics or anything else, and it is essentially what Halphen introduced for conics and what Schubert introduced for arbitrary figures, because classes representing conditions are equal if the conditions, when substituted for each other in any enumerative problem with a finite number of solutions, determine the same number of solutions. For example the classes $\mathbf{z}_i$ are all equal, as all smooth conics are projectively equivalent.

Before algebraic intersection theory had been sufficiently developed for all applications in enumerative geometry, van der Waerden [46 (1930)] pointed out that the topological intersection theory of Solomon Lefschetz (1924, 1926) could be used instead. Severi [36 (1912)], [37 (1916)] had already put the matter definitively in terms of varieties like B and the Z_i; and he had shown that many cases, including most of Chasles's theory, are covered by his algebraic intersection theory, which, however, is at best limited to intersections with cycles of codimension 1. Curiously, it turns out that the cohomology ring of B and its intersection ring modulo numerical equivalence and its intersection ring modulo any one of the other natural equivalence relations all coincide; moreover, the corresponding rings coincide in virtually every important case in enumerative geometry.

The ring $A(B)$ is graded by codimension, and the homogeneous component of degree one $A^1(B)$ has 4 distinguished elements. They are $\mathbf{m}$ and $\mathbf{n}$, the pullbacks of the hyperplane classes in $A(\mathbf{P}^5)$ and $A(\check{\mathbf{P}}^5)$, and $\mathbf{l}$ and $\mathbf{p}$, the classes of the exceptional divisors of $p: B \to \mathbf{P}^5$ and $q: B \to \check{\mathbf{P}}^5$. These classes represent the following conditions, respectively: "to pass through a point"; "to be tangent to a line"; "to be a double line with arbitrary foci"; and "to be a line-pair or be a double line with coincident foci." (The "foci" of a double line necessarily lie on the line.)

The 4 classes are related by the following identities, obtained by Chasles [3, p. 1173 (1864)]:

$$\mathbf{l} = 2\mathbf{m} - \mathbf{n} \quad \text{and} \quad \mathbf{p} = 2\mathbf{n} - \mathbf{m}.$$

The identities were, at the beginning, given geometric proofs ultimately based on the correspondence principle. While these are pretty and can be made rigorous with some effort, it is not worth it; for the identities are special cases of a general result about blowups, which is easy to prove. (The result concerns the general member D of the linear system defining the blowup. It asserts the following: (1) the total transform of D is equal to the proper transform of D plus the exceptional divisor, taken with multiplicity 1; and (2) the proper transform of D is equal to the pullback of a hyperplane under the second projection of the blowup, viewed as the closure of a graph.)

The group $A^1(B)$ is free and has 3 distinguished bases; they are the pairs,

$$(\mathbf{l}, \mathbf{m}) \quad \text{and} \quad (\mathbf{p}, \mathbf{n}) \quad \text{and} \quad (\mathbf{m}, \mathbf{n}).$$

That $(\mathbf{l}, \mathbf{m})$ and $(\mathbf{p}, \mathbf{n})$ are bases follows from the general structure theorem for the intersection ring of a blowup, but it is evident directly. That $(\mathbf{m}, \mathbf{n})$ is a basis follows by virtue of the above identities.

A typical element $\mathbf{z}$ of $A^1(B)$ can be expressed uniquely in the form,

$$\mathbf{z} = \alpha\mathbf{m} + \beta\mathbf{n}$$

for suitable integers α, β. So for any element $\mathbf{y}$ of $A^4(B)$ the intersection number $|\mathbf{z} \cdot \mathbf{y}|$ is given by the formula,

$$|\mathbf{z} \cdot \mathbf{y}| = \alpha\mu + \beta\nu \quad \text{with} \quad \mu = |\mathbf{m} \cdot \mathbf{y}| \quad \text{and} \quad \nu = |\mathbf{n} \cdot \mathbf{y}|.$$

If $\mathbf{z}$ represents a condition on complete conics and $\mathbf{y}$ represents a 1-parameter family of them, then $|\mathbf{z}\cdot\mathbf{y}|$ is the number of members of the family, weighted with the appropriate multiplicities, that satisfy the condition provided the number is finite, and μ and ν are the characteristics of the family, that is, the (weighted) number of members through a point and tangent to a line. Thus Chasles's primary discovery is vindicated.

By the same token, the (weighted) number of complete conics satisfying 5 simple conditions, if finite, is given by the expression,

$$|(\alpha\mathbf{m}+\beta\mathbf{n})(\alpha'\mathbf{m}+\beta'\mathbf{n})(\alpha''\mathbf{m}+\beta''\mathbf{n})(\alpha'''\mathbf{m}+\beta'''\mathbf{n})(\alpha^{\mathrm{iv}}\mathbf{m}+\beta^{\mathrm{iv}}\mathbf{n})|$$

where the α's and β's are the characteristics of the conditions. Since $|m^r n^s|$ is the number of complete conics through r points and tangent to s lines, Halphen's symbolic multiplication is justified.

The 2 identities relating $\mathbf{m}, \mathbf{n}, \mathbf{l}, \mathbf{p}$ were introduced by Chasles [3, p. 1172 (1864)] as the formulas,

$$\lambda = 2\mu - \nu \quad \text{and} \quad \pi = 2\nu - \mu,$$

for the weighted number λ of complete conics in a 1-parameter family with characteristics μ, ν that are double lines with arbitrary foci and for the weighted number π of them that are line-pairs or double lines with coincident foci. Right away Cremona [9 (1864)] used the formulas (and some results of Steiner's) to obtain μ and ν in some cases. Shortly afterwards, but independently, Hieronymus Zeuthen [49 (1865)] developed a general method of obtaining μ and ν via λ and π, including rules for assigning multiplicities to the conics counting in λ and π, and he applied the method to problems involving all sorts of conditions of simple, multiple, and higher order contact. (Zeuthen went to study under Chasles in 1863, but in 1864 he returned home to Denmark to serve in the short war (February–October) against Prussia and Austria.)

Zeuthen's first example [49, II, pp. 21–24, trans. pp. 242–246 (1886)] was the determination of the characteristics of the 5 elementary systems, or of what amounts to the same, the 6 numbers $|\mathbf{m}^r\mathbf{n}^{5-r}|$. (Schubert called numbers of this sort "fundamental numbers.") Here is an adaptation of what Zeuthen did. He started with $|\mathbf{m}^5| = 1$, which obviously holds in view of the definition of $\mathbf{m}$. (It is not so easy to establish the existence of a conic through 5 given points by constructing one in the manner of the ancient Greeks. Isaac Newton considers the matter in the *Principia* (Prop. XXII, Problem XIV, Book I (1687)). He gives the impression he may be the first to do so, but as Prof. Olaf Schmidt of Copenhagen University has pointed out (unpublished), he implicitly assumes the conic exists when justifying its construction (see especially the proof of Lemma XVIII).)

No double line can pass through 3 noncollinear points; whence $|\mathbf{m}^4\mathbf{l}|$ and $|\mathbf{m}^3\mathbf{n}\mathbf{l}|$ are both 0. Therefore we have

$$|\mathbf{m}^4\mathbf{n}| = |\mathbf{m}^4(2\mathbf{m} - \mathbf{l})| = 2|\mathbf{m}^5| - |\mathbf{m}^4\mathbf{l}| = 2,$$

$$|\mathbf{m}^3\mathbf{n}^2| = |\mathbf{m}^3\mathbf{n}(2\mathbf{m} - \mathbf{l})| = 2|\mathbf{m}^4\mathbf{n}| - |\mathbf{m}^3\mathbf{n}\mathbf{l}| = 4.$$

The remaining 3 numbers follow because $|\mathbf{m}^{5-r}\mathbf{n}^r|$ and $|\mathbf{m}^r\mathbf{n}^{5-r}|$ are equal by duality.

Study [41, p. 93 (1886)] discovered and Zeuthen [52, §§168, 171 (1914)] later illustrated how degenerate complete conics could be used to determine the characteristics α, β of a condition: α is the number of line-pairs, consisting of a given line and a line through a given point on the given line, that satisfy the condition; dually β is the number of double lines, with given underlying line and one given focus, that satisfy the condition. For example, consider the condition of tangency to a given smooth conic; α is 2 because there are 2 lines through a given point tangent to the conic, hence 2 line-pairs consisting of a given line and a line through a given point on it that are tangent to the conic, and β is 2 because there are 2 points common to a double line and the conic, hence 2 possibilities for the variable focus. Study gave, in effect, explicit representatives for the 2 classes forming the basis of $A^4(B)$ dual under the intersection pairing to the basis $(\mathbf{m}, \mathbf{n})$ of $A^1(B)$.

The objection to 6^5 can still be raised against 3264: it has not been shown to have enumerative significance! No more than a rigorous proof of Bezout's theorem gives 6^5 significance does a rigorous discussion of $A(B)$ give 3264 significance. Is the number of complete conics tangent to 5 others ever finite? If so, then 3264 is an upper bound, possibly unattained. But are there in general exactly 3264 such conics (in other words, are all the multiplicities 1), and are all 3264 smooth?

Questions of this sort were little considered until a few years ago. Classically, only Halphen seems to have broached them formally. He (see in particular [18, I, pp. 58–61 (1878)]) modified the given conditions independently by applying linear transformations to $\mathbf{P}^2$. For example, modifying the condition of tangency to a conic gives the condition of tangency to the translated conic. Halphen termed conditions "independent" when they are modifications via transformations satisfying no special demand. He was concerned about the smoothness of the conics satisfying independent conditions but not especially about their multiplicities of appearance. After intersection theory had begun to be developed, van der Waerden [47 (1938)] and Severi [37 (1916)], [38 (1940)] spent their efforts primarily on determining the intersection ring of B. They did not compute the number 3264 or any others like it, and they were not especially concerned even in theory about the meaning of such numbers. (Moreover, some of Severi's theoretical treatment, particularly the central lemma of §37, is dubious according to J. G. Semple, writing in some unpublished notes on Chasles's theory. The notes are found in the University of London library. They are written basically in the spirit of Severi's and van der Waerden's works and do not deal with the questions above.)

The questions above can be answered affirmatively with rigor using an elementary result from the general theory of group actions (this remark was made in [24 (1974)]). Indeed, the action of the general linear group on the plane induces one on B. Reformulating the questions in terms of the subvariety Z of B parametrizing the complete conics tangent to a smooth one, we are led to ask if the intersection of 5 independent translates of Z consists of a finite number of points, each appearing with multiplicity 1 and each representing a smooth conic. Now it is not hard to prove (see [23, (4) (1974)]) that, whenever an algebraic group acts transitively on a variety, the intersection of one subvariety and a general translate of another is proper and each component of the intersection appears with multiplicity 1. However, the general linear group has 4 orbits on B; namely, it acts transitively on the sets of smooth conics, of line-pairs, of double lines with distinct foci, and of double lines with coincident foci. Nevertheless, Z does not contain any one of these 4 orbits. It follows that the questions have affirmative answers. (William Hodge and Daniel Pedoe [22, XIV, 7, pp. 359–367 (1952)] answered similar questions about linear spaces satisfying given conditions in a similar way using the (transitive) action of the general linear group on a Grassmann manifold; their mathematics had a hole, which was filled by [23 (1974)].)

The ground field so far has been the complex numbers, but everything considered works the same way over an arbitrary algebraically closed field of characteristic not 2, except for what concerns the multiplicities. The theory of group actions yields in characteristic $p > 2$ only that tangent to 5 general conics there are $(3264/p^e)$ smooth conics each appearing with multiplicity p^e for some $e \geqslant 0$. Thus all the multiplicities are 1, except possibly in characteristics 3 and 17. An ad hoc method shows all the multiplicities are 1 in any characteristic except 2; a direct computation of the tangent space at a general point of the subvariety of B parametrizing the complete conics tangent to a smooth one reveals when 5 such subvarieties will meet transversely.

In characteristic 2 Israel Vainsencher [48 (1975)] proved that there are only 51 conics tangent to 5 general others, and all 51 appear with multiplicity 1 and are smooth. It is a well-known but astonishing fact that in characteristic 2 all the tangents to a smooth conic pass through a common point, the so-called "strange point." Vainsencher showed that the correspondence assigning the strange point to the conic can replace the duality correspondence, and the closure of its graph can replace B. Curiously, it is necessary to revert to the theory of group actions to show the multiplicities are 1 (luckily 51 is odd), for the method used to compute the tangent spaces fails in characteristic 2.

Multiple conditions that do not decompose into several simple conditions were also considered classically. For example a basic double condition of this sort is "to be tangent to a line at a point on it." Chasles [4, p. 349 (1864)] discovered a striking relation involving the class $\mathbf{s}$ in $A^2(B)$ representing this condition: $2\mathbf{s} = \mathbf{mn}$. The relation says, in other words, that the intersection of the variety of complete conics through a point and the variety of complete conics tangent to a line *containing* this point is the variety of complete conics tangent to this line at the point counted twice.

Cremona [9 (1864)] found that the number of complete conics satisfying a double condition and a triple condition independent of it can be expressed in the form,

$$a\rho + b\sigma + c\tau,$$

where a, b, and c depend only on the double condition, and where ρ, σ, and τ denote the numbers of complete conics satisfying the triple condition and, respectively, the following composite conditions: "to pass through 2 points"; "to pass through a point and be tangent to a line"; "to be tangent to 2 lines." Rational numbers were freely used classically, so in modern terms the result says that $\mathbf{m}^2, \mathbf{mn}, \mathbf{n}^2$ form a basis of $A^2(B) \otimes \mathbf{Q}$. In fact, rational numbers are necessary because of the relation $2\mathbf{s} = \mathbf{mn}$.

Cremona's result was implicit in Chasles's work [4, pp. 345–357 (1864)], as Cremona himself says. But Cremona made it explicit, and the expression $a\rho + b\sigma + c\tau$ became known as the *Cremona characteristics formula*. In fact, Cremona dealt only with the case in which the triple condition decomposes into a simple condition and a double condition. However, Halphen [13 (1873)], Lindemann [25, pp. 403–406 (1876)], Schubert and Hurwitz [33 (1876)], Study [42 (1886)], van der Waerden [47 (1938)], and Severi [38 (1940)] all proved Cremona's result when both conditions are indecomposable. Study, van der Waerden, and Severi also delineated the use of rational numbers; they proved that $\mathbf{m}^2, \mathbf{s}, \mathbf{n}^2$ form a basis of $A^2(B)$. This strong form of Cremona's result follows easily from the general structure theorem for the intersection ring of a blowup. (The general structure theorem was discovered by Alexander Grothendieck; see SGA 6, XIV (4.8), p. 677 [*Springer Lecture Notes in Math.*, No. 225 (1971)]. His conjectured "formula clef" was proved by J.-P. Jouanolou [*Invent. Math.*, **11** (1970), 15–26] and by Alex Lascu, David Mumford, and D. B. Scott [*Proc. Cambridge Philos. Soc.*, **78** (1975), 117–123].)

Halphen took a critical second look at his treatment of Chasles's theory, after objections were raised against those of Clebsch and Lindemann. To his great surprise he discovered that Chasles's expression $\alpha\mu + \beta\nu$ is not universally applicable, as everyone had thought. Halphen [15 (1876)], [16 (1876)], [17 (1878)], [18 (1878)], [19 (1879)] found subtle conditions that are satisfied by every double line with coincident foci. (So these degenerate complete conics are sometimes named after him.) Thus 5 such conditions will always have an infinite number of solutions in complete conics, and the number of solutions purported by Chasles's theory has no enumerative significance. To rectify the situation Halphen introduced additional characteristics and replaced Chasles's expression by

$$a\mu + b\nu + c\rho + c'\rho' + \cdots . \qquad (**)$$

The number of nonzero terms in any given case is of course finite, but for any given n there is an enumerative problem requiring n nonzero terms. In effect, Halphen blew up the variety B of complete conics along the subvariety of double lines with coincident foci, then he blew up again and again as often as necessary for a given problem.

Halphen, however, did not focus in on the case of an infinite number of solutions. His examples of 1876 and of 1878 and virtually all those of 1879 had finitely many solutions. Rather, he was primarily concerned about another matter, which these examples illustrate: even though given conditions are fully independent, and each is satisfied by some smooth conic, and together they determine a finite number of solutions, nevertheless some solutions can be degenerate and the degenerate ones are necessarily double lines with coincident foci. Halphen discounted these degenerate solutions; he seems to take it for granted that they are unacceptable. In his important article [18, I (1878)] he gives an a priori basis for deciding which solutions to count; namely, count only those that vary when any one of the conditions is changed by the application of a projective transformation of the plane. Study [41, pp. 62–63 (1886)] later termed these "movable solutions." Halphen's formula (**) with the additional characteristics is for the number of movable solutions.

Halphen's use of additional characteristics can be justified using the theory of group actions and the following fact: for each subvariety D of B with codimension 1 that is not stable under the action of the general linear group, there exists a variety B_1 obtained from B via a sequence of equivariant blowups with smooth centers such that the strict transform of D on B_1 meets each orbit of B_1 properly. Heisuke Hironaka (private communication) suggests obtaining B_1 as follows: first blow up B along the scheme-theoretic intersection of all the translates of D; then, using his general theory, dominate this blowup by a variety B_1 obtained from B via a sequence of equivariant blowups with smooth centers. The ground field must have characteristic zero here both for the theory of group actions and for Hironaka's theory. (Some work on the case of one additional characteristic was done from a slightly different point of view by Severi [38, §§52–53 (1940)] and J. G. Semple [*J. London Math. Soc.*, **26** (1951), 122–125]. Semple also suggests considering a sequence of blowups. Zeuthen [52, §§169–170 (1914)] explains how to find the number of movable solutions without using any additional characteristics by carefully considering the path of degeneration in the 1-parameter family.)

Halphen's discoveries raised a famous controversy over the philosophical basis of Chasles's theory. Right after the appearance of Halphen's note [15 (1876)], Schubert and Hurwitz published their treatment [33 (1876)] of Chasles's theory. In it they said that Halphen's examples only underscore the need for an appropriate interpretation of Chasles's formula $\alpha\mu + \beta\nu$, and they proved that the formula is valid if all (finitely many) solutions in complete conics are admitted. Later, convinced by Halphen's full-length article [18 (1878)], Schubert [30 (1880)] (in a rare use of French) repudiated the work for the case that some double lines with coincident foci appear as solutions. (He referred to the version in his book [29, §38, pp. 284–288 (1879)] and thus left Hurwitz's name out of it.)

What especially persuaded Schubert was Halphen's discussion of degenerate conics. Here and elsewhere one senses that many classical geometers had a platonic view of figures like conics. There are ideal conics of which we see only shadows or aspects like their point-sets and their envelopes of tangent lines. Smooth conics are determined by their points alone and by their tangents alone. Line-pairs and double lines with distinct foci require both. Higher degenerate forms require additional aspects to completely determine them. Such higher forms were discovered by Halphen, although he did not make that fact clear before his full-length article [18 (1878)], and they came as a surprise to Schubert in particular.

Study later, in his treatment [41 (1886)] of Chasles's formula $\alpha\mu + \beta\nu$, took the original point of view of Schubert and Hurwitz (without saying it was theirs), that the formula could be saved. Study's article was reviewed straight off by Schubert [31 (1886)], who seems to revert back to this position, for he does not criticize it or even mention Halphen's work on degenerate forms and additional characteristics. (Also he does not mention his own encounter with the matter.) Halphen earlier, before Schubert's disclaimer appeared, expressed his view of this philosophy of salvage, which he felt he had found several places in Schubert's book. In a letter [20 (1879)] to Zeuthen (who apparently was a close friend of his) Halphen wrote, "M. Schubert veut absolument changer la nature pour l'accommoder à ses formules." He also wrote that the replacement (**) for $\alpha\mu + \beta\nu$ was a formula that "j'ai eu tant de peine à trouver et que je défendrai jusqu'à la mort *unquibus et rostro*." Halphen died prematurely in 1889, and Felix Klein, who edited the *Mathematische Annalen* at the time, asked Zeuthen to comment publicly on Study's article. Zeuthen published some criticism [50 (1890)], which he extracted from a letter of his to Klein. Study [44 (1892)], who felt sharply attacked, rebutted it, restating his position. Zeuthen [51 (1893)] responded trying to further clarify his, and Halphen's, position. After that Severi [37 (1916)], [38 (1940)] accepted Halphen's view, while van der Waerden [47 (1938)] accepted Study's.

Zeuthen said essentially this: When Chasles's formula $\alpha\mu + \beta\nu$

is saved by admitting all solutions in complete conics, the accepted meaning of the problem being solved is changed; in other words, some of the admitted solutions are extraneous, introduced simply by the method of solution. Consider de Jonquières's theory, for example. It suggests there are a finite number, 8, of conics through 2 general points and tangent to 3 general lines; however, the (ordinary) double line through the 2 points, which appears with multiplicity 4, is an extraneous solution. (Zeuthen did not mention any specific example of de Jonquières theory; this particular example was discussed by Cremona [8, 2, (1864)].) In this case and in many others Chasles's theory is better because it introduces no extraneous solutions. Best of all is Halphen's theory because it never introduces extraneous solutions.

Chasles won the Copley medal of the Royal Society of London in 1865 particularly for his work on characteristics. On awarding the medal the society's president General Edward Sabine [28 (1865)] said, "... it is probable that, as an instrument of purely geometrical research, the method of Chasles will bear comparison with any other discovery of the century." In fact, virtually from the beginning Chasles's theory of conics was generalized to quadratic surfaces and conics in 3-space by de Jonquières, Chasles himself, Zeuthen, Schubert, Halphen, and others. (A guide to the literature is given by Zeuthen-Pieri [53, pp. 305–307, 315–316 (1915)] and by Segre [34, pp. 855–856 (1912)].) For example, Schubert shows in his book [29, p. 106 (1879)] how to find the number 666,841,088 of quadric surfaces tangent to 12 given quadric surfaces. Eventually Schubert [32 (1894)] did the definitive work, treating quadric q-folds in n-space. Making this work rigorous is today not a completely straightforward matter—some new ideas are necessary—but it may not be far beyond our reach.

REFERENCES

1. Johann Nikolai Bischoff, "*Einige Sätze über die Tangenten algebraischer Curven*", Crelle's J. Reine Angew. Math. [56 (1859), 166–177].
2. William Braikenridge, 1700–1759, *Exercitatio geometrica de descriptione linearum curvarum*, London (1733).

3. Michel Chasles, 1793–1880, Series of 4 notes in C. R. Acd. Sci. Paris t.58 (1864): "*Détermination du nombre des sections coniques qui doivent toucher cinq courbes données d'ordre quelconque, ou satisfaire à diverses autres conditions*" [1 fév., 222–226]; "*Construction des coniques qui satisfont à cinq conditions. Nombres des solutions dans chaque question*" [15 fév., 297–308]; "*Systèmes de coniques qui coupent des coniques données sous des angles donnés, ou sous des angles indéterminés, mais dont les bissectrices ont des directions données*" [7 mars, 425–431]; *Considérations sur la méthode générale exposée dans la séance du 15 février.—Différences entre cette méthode et la méthode analytique—Procédés généraux de démonstration*" [27 juin, 1167–1175].
4. ———, Series of 4 notes in C. R. Acd. Sci. Paris t.59 (1864): "*Exemples des procédés de démonstration annoncés dans la séance précédente*" [4 juillet, 7–15]; "*Suite des propriétés relatives aux systèmes de sections coniques*" [18 juillet, 93–97]; "*Questions dans lesquelles il y a lieu de tenir compte des points singuliers des courbes d'ordre supérieur.—Formules générales comprenant la solution de toutes les questions relatives aux sections coniques*" [1 août, 209–218]; "*Questions dans lesquelles entrent des conditions multiples, telles que des conditions de double contact ou de contact d'ordre supérieur*". [22 août, 345–357].
5. ———, "*Observations relatives à la théorie des systèmes de courbes*", C. R. Acd. Sci. Paris [63 (1866), 816–821].
6. ———, Series of 4 notes in C. R. Acd. Sci. Paris t.72 (1871): "*Propriétés des systèmes de coniques, relatives, toutes, à certaines séries de normales en rapport avec d'autres lignes ou divers points*" [10 avril, 419–430]; "*Propriétés des systèmes de coniques, dans lesquels se trouvent des conditions de perpendicularité entre diverses séries de droites*" [24 avril, 487–494]: "*Théorèmes divers concernant les systèmes de Coniques représentés par deux caractéristiques*" [1 mai, 511–520]; "*Propriétés des courbes d'ordre et de classe quelconques démontrées par le principe de correspondance*" [15 mai, 577–588].
7. Alfred Rudolf Friedrich Clebsch, 1833–1872, "*Zur Theorie der Charakteristiken*", Math. Ann. [6 (1873), 1–15].
8. Antonio Luigi Gaudenzio Giuseppe Cremona, 1830–1903, "*Sulla teoria delle coniche*", Giorn. Mat. [(1) II (1864), 17–20]. *Opere Matematiche*, t.II, Ulrico Hoepli, Milano (1915), 95–99.
9. ———, "*Sur le nombre des coniques qui satisfont à des conditions doubles*", C. R. Acd. Sci. Paris [59 (1864), 776–779]. *Opere Matematiche*, t.II, Ulrico Hoepli, Milano (1915), 119–122.
10. Jean Philippe Ernest de Faque de Jonquières, 1820–1901, "*Théorèmes généraux concernant les courbes géométriques planes d'un ordre quelconque*", Liouville's J. Math. Pures Appl. [(2) 6 (1861), 113–134].
11. ———, "*Sur la détermination des valeurs des caractéristiques dans les séries ou systèmes élémentaires de courbes et de surfaces*", C. R. Acd. Sci. [63 (1866), 793–797].

12. ———, "*Mémoire sur les contacts multiples d'ordre quelconque des courbes de degré r, qui satisfont à des conditions données, avec une courbe fixe du degré m; suivi de quelques réflexions sur la solution d'un grand nombre de questions concernant les propriétés projectives des courbes et des surfaces algébriques*", Crelle's J. Reine Agnew. Math. [66 (1866), 289–321].
13. Georges-Henri Halphen, 1844 – 1889, "*Mémoire sur la détermination des coniques et des surfaces du second ordre*", Bull. Soc. Math. France [I (1872–73) 130–148, 226–240; II (1873–74), 11–33]. Oeuvres, t.I, Gauthier-Villars, Paris (1916), 98–157.
14. ———, "*Sur les caractéristiques dans la théorie des coniques sur le plan et dans l'espace, et des surfaces du second ordre*", C. R. Acd. Sci. Paris [76 (1873), 1074–1077]. *Oeuvres*, t.I, Gauthier-Villars, Paris (1916), 159–162.
15. ———, "*Sur les caractéristiques des systèmes des coniques*", C. R. Acd. Sci. Paris [83 (1876), 537–539], *Oeuvres*, t.I, Gauthier-Villars, Paris (1916), 543–545.
16. ———, "*Sur les caractéristiques des systèmes de coniques et de surfaces du second ordre*", C. R. Acd. Sci. Paris [83 (1876), 886–889]. *Oeuvres*, t.I, Gauthier-Villars, Paris (1916), 553–556.
17. ———, "*Sur les caractéristiques des systèmes de coniques et de surfaces du second ordre*", Journal de l'Ecole Polytechnique Paris [45e cahier, 28 (1878), 27–84]. *Oeuvres*, t.II, Gauthier-Villars, Paris (1918), 1–57.
18. ———, "*Sur la théorie des caractéristiques pour les coniques*", Proc. Lond. Math. Soc. [9 (1877–78), 149–177], reprinted in Math. Ann. [15 (1879), 16–44]. *Oeuvres*, t.II, Gauthier-Villars, Paris (1918), 58–92.
19. ———, "*Sur le nombre des coniques qui, dans un plan, satisfont à cinq conditions projectives et indépendantes entre elles*", Proc. Lond. Math. Soc. [10 (1878–79), 76–91]. *Oeuvres*, t.II, Gauthier-Villars, Paris (1918), 275–289.
20. ———, *Extract of a letter to Zeuthen dated 7 December 1879*, *Oeuvres*, t.IV, Gauthier-Villars, Paris (1924), 636–637.
21. David Hilbert, 1862–1943, "*Mathematical problems*", translated by Dr. Mary Winston Newson, Bull. Amer. Math. Soc. [50 (1902), 437–479].
22. William Vallance Douglas Hodge, (1903–), and Daniel Pedoe, *Methods of Algebraic Geometry*, Cambridge University Press, vol. II (1952), reprinted 1968.
23. Steven Lawrence Kleiman, (1942–), "*The Transversality of a general translate*", Compositio Math. [(3) 28 (1974), 287–297].
24. ———, "*Problem 15. Rigorous foundation of Schubert's enumerative calculus*", Amer Math. Soc. Symposium on Hilbert's problems, De Kalb (1974), AMS Series PSPM 28.
25. Carl Louis Ferdinand Lindemann, 1852–1939, in *Vorlesungen über Geometrie* von A. Clebsch, ersten Bandes, *Geometrie der Ebene*, Leipzig, Teubner (1876), 390–425.
26. Eugène Prounet, (?–1867), "*Sur le nombre des coniques qui satisfont à cinq conditions données, d'après M. Chasles*", Nouv. Ann. Math. Paris [(2) 5 (1866), 193–203].

27. ——, "Note du rédacteur", Nouv. Ann. Math. Paris [(2) 5 (1866), 211–213].
28. Edward Sabine, "President's Address", Proc. Royal Society London [14 (1865), 493–496], transl. Nouv. Ann. Math. Paris [(2) V (1866), 84–91].
29. Herman Cäser Hannibal Schubert, 1848–1911, *Kalkül der abzählenden Geometrie*, Teubner, Leipzig (1879).
30. ——, "*Note sur l'évaluation du nombre des coniques faisant partie d'un système et satisfaisant à une condition simple*", Bull. Soc. Math. France [8 (1879/80), 61].
31. ——, Review of Study's "Ueber die Geometrie der Kegelschnitte, insbesondere deren Charakteristikenproblem", Jahrbuch über die Fortschritte der Math. [18 (1886), 629–630].
32. ——, "*Allgemeine Anzahlfunctionen für Kegelschnitte, Flächen und Räume zweiten Grades in n Dimensionen*", Math. Annalen [45 (1894), 153–206].
33. H. Schubert and Adolf Hurwitz, 1859–1919, "*Über den Chasles'schen Satz* $\alpha\mu + \beta\nu$", Nachrichten von der k. Gesellschaft der Wissenschaften zu Göttingen [(1876), 503–517]. *Mathematische Werke* von Hurwitz, Band II, E. Birkhäuser & Cie., Basel (1933).
34. Corrado Segre, 1863–1924, "*Mehrdimensionale Räume*", Encyclopadie der Mathematischen Wissenschaften mit Einschluss ihrer Anwendungen, [III, 2, 2, C7; 669–972], (Abgeschlossen Ende 1912), B. G. Teubner, Leipzig, (1921–1934).
35. Francesco Severi, 1879–1963, "*Sulle intersezioni delle varietà algebriche e sopra i loro caratteri e singolarità proiettive*", Memorie della R. Accademia delle Scienze di Torino [52 (1902), 61–118].
36. ——, "*Sul principio della conservazione del numero*", Rendiconti del Circolo Matematico di Palermo [33 (1912), 313–327].
37. ——, "*Sui fondamenti della Geometria numerativa e sulla teoria delle caratteristiche*", Atti del R. Instituto Veneto [75 (1916), 1121–1162].
38. ——, "*I fondamenti della geometria numerativa*", Annali di Mat. [4, 19 (1940), 153–242]. Also available in a German translation by Wolfgang Gröbner, entitled "Grundlagen der abzählenden Geometrie" published in 1948 by Wolfenbutteler Verlagsanstalt G.M.B.H., Wolfenbüttel and Hannover.
39. Jacob Steiner, 1796–1860, "*Elementare Lösung einer geometrischen Aufgabe, und über einige damit in Beziehung stehende Eigenschaften der Kegelschnitte*", Crelle's J. Reine Angew. Math. [37 (1848), 161–192]. (Auszug aus einer am 19ten April 1847 der Akademie der Wissenschaften vorgelegten Abhandlung.) Gesammelte Werke, Band II, Reimer, Berlin (1882), 389–420.
40. ——, "*Allgemeine Eigenschaften der algebraischen Curven*", Crelle's J. Reine Angew. Math. [47 (1854), 1–6] (Abgedruckt aus dem Monatsbericht der hiesigen Akad. der Wissens. von August 1848). Gesammelte Werke, Band II, Reiner, Berlin (1882), 493–509.
41. Eduard Study, 1862–1922, "*Ueber die Geometrie der Kegelschnitte, insbesonderen deres Charakteristikenproblem*", Math. Annalen [27 (1886), 58–101].

42. ———, "*Ueber die Cremona'sche Charakteristikenformel*", Math. Ann. [27 (1886), 102–105].
43. ———, "*Abbildung der Mannigfaltigkeit aller Kegelschnitte einer Ebene auf einen Punktraum*", Math. Ann. [40 (1892), 551–558].
44. ———, "*Entgegnung*", Math. Ann. [40 (1892), 559–562].
45. ———, "*Über Systeme von Kegelschnitten*", Math. Ann. [40 (1892), 563–578].
46. Batel Leendert van der Waerden, 1903–, "*Topologische Begründung des Kalküls der abzählenden Geometrie*", Math. Annalen [102 (1930), 337–362].
47. ———, "*Zur algebraischen Geometrie. XV. Lösung des Charakteristikenproblems für Kegelschnitte*", Math. Annalen [115 (1938), 645–655].
48. Israel Vainsencher, 1948–, "*Conics in Characteristic 2*", Master's thesis, M.I.T., Cambridge, MA, first draft December 1975, Compositio Math. [36, 1 (1978), 101–112].
49. Hieronymus Georg Zeuthen, 1839–1920, "*Nyt Bidrag til Laeren om Systemer af Keglesnit, der ere underkastede* 4 *Betingelser*". Diss. Copenhagen (1865). Transl. entitled, "Nouvelle méthode pour déterminer les caractéristiques des systèmes de coniques", Nouv. Ann. Math. Paris [(2) 5 (1866), 241–62, 289–97, 385–98, 433–43, 481–92, 529–40].
50. ———, "*Sur la révision de la théorie des caractéristiques de M. Study*", Math. Ann. [37 (1890), 461–464].
51. ———, "*Exemples de la détermination des coniques dans un système donné qui satisfont à une condition donnée*", Math. Ann. [41 (1893), 539–544].
52. ———, *Lehrbuch der abzählenden Methoden der Geometrie*, Teubner, Leipzig (1914).
53. H. G. Zeuthen and M. Pieri, "*Géométrie énumérative*", Encyclopédie des sciences mathématiques, [III, 2, 260–331] Teubner, Leipzig (1915).

INDICES
(BY AUTHOR)

Maxwell Rosenlicht, "Linear Algebraic Groups" (pages 1–18)

B. L. van der Waerden, "The Connectedness Theorem and Multiplicity" (pages 19–45)

Jack Ohm, "Space Curves" (pages 47–115)

List of symbols used
(in order of appearance—pages 47–115)

Steven L. Kleiman, "Chasles's Enumerative Theory of Conics" (pages 117–138)